The Economics of Electricity Markets

LOYOLA DE PALACIO SERIES ON EUROPEAN ENERGY POLICY

Series Editor: Jean-Michel Glachant, *Loyola de Palacio Professor for European Energy Policy and Director of the Florence School of Regulation, European University Institute, Italy, and Professor of Economics, Université Paris-Sud, France*

The *Loyola de Palacio Series on European Energy Policy* honours Loyola de Palacio (1950–2006), former Vice-President of the European Commission and EU Commissioner for Energy and Transport (1999–2004), a pioneer in the creation of an EU Energy Policy.

This series aims to promote energy policy research, develop academic knowledge and nurture the 'market for ideas' in the field of energy policy making. It will offer informed and up-to-date analysis on key European energy policy issues (from market building to security of supply; from climate change to a low carbon economy and society). It will engage in a fruitful dialogue between academics (including economists, lawyers, engineers and political scientists), practitioners and decision-makers. The series will complement the large range of activities performed at the Loyola de Palacio Chair currently held by Professor Jean-Michel Glachant (Robert Schuman Center for Advanced Studies at European University Institute in Florence, Italy, and University Paris-Sud 11).

Titles in the series include:

Security of Energy Supply in Europe
Natural Gas, Nuclear and Hydrogen
Edited by François Lévêque, Jean-Michel Glachant, Julián Barquín, Christian von Hirschhausen, Franziska Holz, and William J. Nuttall

Competition, Contracts and Electricity Markets
A New Perspective
Edited by Jean-Michel Glachant and Dominique Finon

The Economics of Electricity Markets
Theory and Policy
Edited by Pippo Ranci and Guido Cervigni

The Economics of Electricity Markets

Theory and Policy

Edited by

Pippo Ranci

Università Cattolica, Italy

Guido Cervigni

Università Bocconi, Italy

THE LOYOLA DE PALACIO SERIES ON EUROPEAN ENERGY POLICY

Edward Elgar

Cheltenham, UK • Northampton, MA, USA

Published by
Edward Elgar Publishing Limited
The Lypiatts
15 Lansdown Road
Cheltenham
Glos GL50 2JA
UK

Edward Elgar Publishing, Inc.
William Pratt House
9 Dewey Court
Northampton
Massachusetts 01060
USA

This book has been printed on demand to keep the title in print.

A catalogue record for this book
is available from the British Library

Library of Congress Control Number: 2012955225

This book is available electronically in the ElgarOnline.com
Economics Subject Collection, E-ISBN 978 0 85793 396 6

ISBN 978 0 85793 395 9

Typeset by Servis Filmsetting Ltd, Stcokport, Cheshire

Contents

Contributors

Guido Cervigni is Research Director at IEFE, the Center for Research on Energy and Environmental Economics and Policy at Bocconi University, Milan, Italy and Chief Economist at A2A S.p.A. He is an economist and former regulator who has provided qualitative and quantitative analyses on regulatory and competition-related issues, with a particular focus on energy markets. He has advised regulators, system operators, exchanges and businesses across Europe on a wide range of regulatory issues in energy markets, including market design, regulatory frameworks supporting low-carbon technologies, price regulation, contracts and asset evaluation. He has been Director at LECG Consulting in London and Head of LECG's Italian office. Prior to joining LECG, he was Head of Economic Analysis and Regulatory Affairs at Enel S.p.A., Head of Business Development in an energy trading company and Head of Market Development at the Italian Power Exchange. He started his career at the Italian Energy Regulatory Authority where he was Head of the Competition and Markets Division. He holds a PhD in economics from Bocconi University, Milan.

Andrea Commisso is an economist in the Markets Division of the Italian Energy Regulatory Authority (AEEG). The main focus of his work at AEEG is the regulation and monitoring of the Italian wholesale electricity and ancillary service markets. Prior to joining AEEG, he was Associate Economist at LECG Consulting in London, where he specialised in the economic analysis of energy markets. During his years in consultancy he also provided empirical analysis on a variety of antitrust matters outside of the energy industry, including merger analysis, market definition and assessment of market power. He graduated in economics from Bocconi University, Milan and holds an MSc in Competition and Market Regulation from Pompeu Fabra University, Barcelona.

Anna Creti is Professor in the Université Paris Ouest Nanterre la Défense and Deputy Director of the Economic Research Unit, Paris, France. She is senior research associate at École Polytechnique, a member of the Scientific Committee of the French section of the World Energy Council and a member of the Scientific Committee of the EnergyLab Foundation. Prior to joining Université Paris Ouest Nanterre la Défense, she was a researcher at the Institut d'Économie Industrielle and at the Laboratoire d'Économie des Ressources Naturelles, University of Toulouse I and later Associate Professor at Bocconi University, Milan, from 2005 to 2009, where she was also the Research Director at the Center for Research of Energy and Environmental Economics and Policy. Nominated in 2009 for the ENI award, she is the author of numerous articles in international journals such as *Energy Economics, Resource and Energy Economics, Economica* and *Energy Policy*. She also serves as consultant for major energy firms, as well as energy and environmental regulators.

Dmitri Perekhodtsev is a senior economist at FTI Compass Lexecon (formerly LECG Consulting) in Paris, France. His professional practice focuses on economic aspects of regulation and competition cases, in particular in the European energy sector. He has advised European national energy regulators, power exchanges, transmission system operators and major energy companies on issues related to analyses of competition and abuse of market power in the wholesale electricity markets, development of efficient design of electricity markets and monitoring their efficiency, developing models of cross-border transmission capacity allocation in the context of electricity market integration initiatives in Continental Europe, and developing regulatory incentive mechanisms for transmission development. Prior to joining LECG in Europe, he was a consultant at LECG's US-based energy group specialising in the competition and design of the US electricity markets. He holds a PhD in economics from Carnegie Mellon University. He has written research papers and co-authored book chapters on various issues related to international electricity markets.

Clara Poletti is Director of International Affairs, Strategy and Planning at Autorità per l'energia elettrica e il gas, the Italian energy regulator and former Director of the Center for Research on Energy and Environmental Economics and Policy (IEFE) at Bocconi University, Milan, Italy. Prior to joining IEFE, she was first an economist in the tariff division, then Head of the Market and Competition Division at Autorità per l'energia elettrica e il gas. She also led the regulator's Market Monitoring Unit, responsible for the monitoring of the Italian wholesale electricity and ancillary service markets. She has a longstanding experience on wholesale markets design, incentive mechanisms and competition, mainly in the electricity sector. She has published in international economic journals, including the *Journal of Industrial Economics*, the *Journal of Economic Theory* and *Economics Letters*.

Pippo Ranci is Professor of Economic Policy at the Catholic University of Milan, Italy, where he currently teaches Ethics of Finance. He chairs the

Board of Appeal of the European Agency for the Cooperation of Energy Regulators. Between 1996 and 2003 he was the first President of the Italian Regulatory Authority for Energy; in that role he was among the founders of the Council of European Energy Regulators. He was the first director of the Florence School of Regulation at the European University Institute, and is currently the President of the Supervisory Board of A2A S.p.A., the second-largest electricity company in Italy. He graduated from the Catholic University, Milan and the University of Michigan.

1. Introduction

Pippo Ranci and Guido Cervigni

1.1 OVERVIEW

Over the last 20 years great efforts have been made in all the advanced economies to introduce the market into the electricity industry, which had traditionally been dominated by a monopoly integrated horizontally – typically on a national scale – and vertically, from production to supply to the end users.

The reasons for this development are many, and exist in different combinations in different countries. First, some of the conditions that justified the adoption of monopolistic models no longer held. Liberalisation policies were typically developed and implemented in a situation of relative maturity of the sector, that is, when the phase of major investments required for provision of electricity to the population as a whole had essentially been completed. This suggests that recipes for liberalisation drawn from the experiences of industrialised countries should not be transposed to developing countries without an in-depth understanding of the specific situation.

Developments in technology also contributed to overcoming the perception that a monopoly in electricity generation was the 'natural' solution. Towards the end of the 1980s, combined-cycle gas turbine generation technology became available. Compared to traditional thermal technologies, combined cycles offered lower installation and management costs, a low environmental impact, and a standardised, modular design. At the same time, developments in information technology and telecommunications allowed low-cost management of the large quantities of information necessary for the operation of the wholesale and retail electricity markets.

Second, liberalisation policies were often motivated by dissatisfaction with the services of the monopoly suppliers and of the institutions responsible for their regulation. The reasons for dissatisfaction ranged from investment gold-plating to poor-quality service. The limitations of the regulatory bodies were identified as a lack of independence from the companies they were regulating, an excessive tendency towards micro-management, and a lack of accountability.

In this perspective, the creation of wholesale electricity markets can be interpreted as removing an anomaly. Electricity began to be considered and traded as a good like many others. Yet, liberalisation policies also reflected genuine ideological motives, such as the conviction that competition and privatisation would deliver superior investment decisions, better quality of service and lower supply costs in the electricity industry, and that the possibility for consumers to choose between several suppliers is intrinsically valuable.

As a result of a series of institutional experiments carried out in North America and the United Kingdom between 1990 and the early years of the twenty first century, reasonably robust and efficient wholesale electricity markets now operate in many countries. By 2010 more than half of the countries in the world had introduced a reform process in their wholesale power sectors, accounting for about 60 per cent of the worldwide electricity production. Liberalisation of the retail segment has been less popular: by 2008 only 23 OECD countries had fully liberalised electricity retailing, accounting for about 20 per cent of the worldwide electricity demand.

An effective market design can make electricity very much like other commodities as far as production, trading and retailing are concerned. Social and political concerns, rather than engineering or economics, still make electricity 'special'. However, the unique technical features of electricity, such as the lack of cheap techniques for storing it, make complex organisational arrangements necessary to support trading and deal with some well-justified competition policy concerns.

This book addresses the main issues arising when competition is introduced in the electricity industry. The selection of topics and perspective of the analysis reflect the background and experience of the authors. These include economists who lived through various moments of the opening of the European electricity markets in their positions as professionals working for governments, system operators, power exchanges and energy businesses, or who were engaged in regulatory activity exposed to the political dimension of the liberalisation process, or indeed were carrying out research.

Our work aims to be broadly useful and applicable. Most references are to stylised settings. However, we also draw extensively from specific arrangements in order to illustrate the practical relevance of the general concepts. The book was conceived as an educational tool for courses on the economics of the electricity industry, but also as a straightforward and accurate guide to be used by industry professionals. The only requirement for reading it is a basic knowledge of economics.

The authors would like to thank Simona Benedettini and Silvia Concettini for precious research assistance. Our special thanks go to our reviser Sally Winch, who ironed out any inconsistencies and language issues.

The remainder of this chapter provides an overview of the topics developed in the following chapters. In Section 1.2 we introduce the basics of electricity supply technology. In Section 1.3 we investigate why power systems need to be run by a central entity, known as a system operator (SO). In Sections 1.4 and 1.5 we discuss how the technical features of electricity shape the arrangements governing the trading and delivery of electricity. In Section 1.6 we address the concern that liberalised electricity markets may not be able to attract adequate investment in generation capacity. In Section 1.7 we focus on the issues related to electricity networks. In Section 1.8 we analyse the market-power issues specifically identified in electricity generation. In Section 1.9 we discuss electricity retailing. Finally, in Section 1.10 we address the changes to the organisation of electricity markets following the measures being implemented to address climate change.

1.2 TECHNICAL AND ECONOMIC FEATURES OF ELECTRICITY SUPPLY

Power is the quantity of electricity produced or consumed at a certain point in time, measured in Watts (W).[1] The quantity of electricity produced or consumed during a certain period of time is measured in Watt-hours (Wh). A Wh is the energy consumed by the continuous application of 1 W of power for one hour. Figure 1.1 provides some basic information about the European electricity industry in 2009, when total European generation capacity was 840 GW and electricity consumption amounted to 2,810 TWh. The order of magnitude of annual consumption by large industrial customers is the GWh. Annual consumption by residential customers is in the range of a few MWh.

Several features combine to make electricity different from other commodities. First, electricity cannot be economically stored on a large scale. It must constantly be produced in the same quantity as it is consumed. Widespread and uncontrolled service interruptions ensue if electricity injections into and withdrawals from the network do not match, even just for seconds.

Second, electricity demand varies significantly during the day and across the seasons. Figure 1.2 shows electricity consumption in Italy on a typical working day. Withdrawals can vary considerably within a day. In

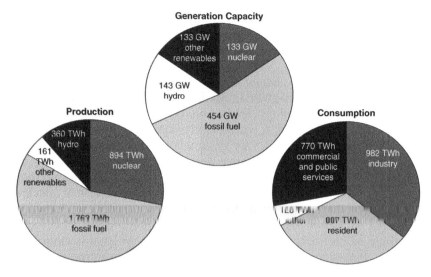

Source: Data International Energy Agency.

Figure 1.1   EU27 generation capacity, production and consumption in 2009

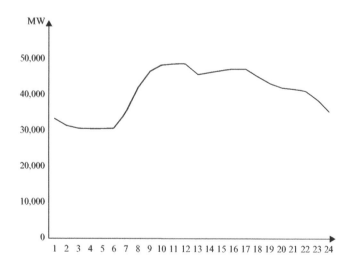

Source: Data Terna S.p.A.

Figure 1.2   Hourly load values for the third Wednesday of June 2009 in Italy

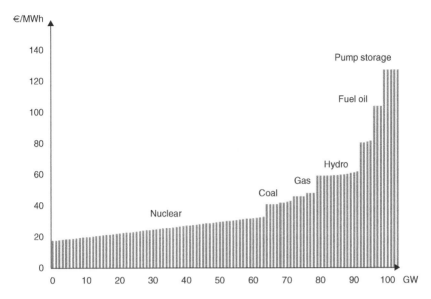

Source: German Federal Ministry for the Environment, Nature Conservation and Nuclear Safety, 'Renewable energy policy in Germany and the European Union', JREF Conference, Tokyo, 9 March 2012.

Figure 1.3   Generation merit-order curve, France, 2009

the day shown in the figure, for example, withdrawals at 11.00 were 60 per cent greater than those at 04.00.

Since electricity must be generated at the same time it is consumed, and demand varies, an efficient generation fleet consists of a mix of different generation technologies with different fixed/variable cost ratios. High fixed-cost and low variable-cost units include run-of-river hydroelectric, nuclear and coal units. These typically produce all the time, and are therefore available to meet the base load. Higher variable-cost units produce when consumption is greater than the available base-load generation capacity. Combined-cycle gas generators typically have intermediate load factors. Old steam generators burning oil or gas are positioned further up in the system variable-cost curve, and are activated when demand is higher than the cheaper available capacity. Finally, high-variable cost open-cycle gas turbines and controllable hydropower generators operate at peak times only.

Figure 1.3 represents the estimated variable-cost curve in France in 2009. The variable-cost curve is also referred to as the generating units' 'merit order'. Since the supply function includes units with different variable costs, the variable cost of the most expensive active generator, or the system marginal cost, fluctuates as demand varies, hitting different segments of the supply function. The system marginal cost may be one order of magnitude greater in peak hours than in low-demand hours.[2] In order to address these large cost differences, in most wholesale electricity markets electricity produced and consumed in different hours, or half-hours, is regarded as a different product, so that the market-clearing price varies from hour to hour.

Third, electricity differs from other commodities because a large portion of the demand for electricity is currently price inflexible, at least in short timeframes. Consumers may not react to spot prices for several reasons. The meters currently operating at most small consumers' premises only record total consumption over a long period. Traditional meters only record total consumption since the meter was first activated, which only allows calculation of the consumer's withdrawal between two readings. More recent meters record total consumption over fixed periods, such as one month. It is generally impossible to assess a consumer's withdrawal per hour or half-hour. Furthermore, even if hourly or half-hourly consumption is recorded, adjusting consumption in response to spot prices could entail high transaction costs for monitoring the spot prices and adapting appliances. In addition, the current arrangements do not allow price-dependent consumption decisions in situations of scarcity, that is, when available generation capacity is insufficient to meet demand. In fact, when scarcity occurs, rationing is implemented at distribution area level and not on a consumer-by-consumer basis.

When hourly consumption is not known, retail prices cannot directly reflect wholesale market prices, and therefore cannot convey to consumers the economic signals of the cost caused by consumption at each time. The consumers can only be charged a price that reflects the estimated average cost caused during the period between two meter readings, based on a conventional time pattern. This practice is known as 'load profiling'. As a consequence, the consumers' demand cannot respond to the high hourly prices that clear the wholesale spot market when generation falls short of demand, because they pay the same price irrespective of when consumption takes place.

Fourth, electricity is delivered to consumers via transmission and distribution networks. Transmission networks allow electricity to be moved long distances from the generators to the consumption areas.[3] Transmission networks are subject to congestion, that is, the power flows corresponding to the electricity market transactions may violate some of the network's security constraints. Congestion is relieved mainly by changing the distribution of total production across the generating units connected at different locations.

Distribution networks are used to transfer electricity from the transmission network to the customers' premises. Distribution networks are typically dimensioned in such a way that congestion does not occur in normal conditions, as congestion on the distribution network could result in service interruptions.

In the following sections we discuss how electricity's technical features affect the market design and industry operations, and refer to the chapter in which each issue is addressed.

1.3   THE SYSTEM OPERATOR

The rationale for a central entity running the power system is undisputed. Without a central system operator, coordinating the production and consumption decisions of all network users to ensure that system security conditions are met at all times would be impossible or prohibitively expensive. Service disruptions caused by failures to meet the system security constraints would result in extremely large welfare losses. For those reasons, placing the responsibility of maintaining security at all times on a system operator is, in practice, the only feasible solution.

System security requires production and consumption to constantly match, power flows not to violate any network constraints, and sufficient spare transmission and generation capacity to be available in order to avoid service interruptions in the event of outages or unexpected surges in demand.

In order to ensure that the power system is balanced and secure at all times, the system operator buys ancillary services from generators and possibly from large consumers with the necessary capabilities. Ancillary services include reserve capacity, the commitment to make generation capacity with certain technical capabilities available at a certain time.

In addition, the system operator operates the real-time or balancing market, where additional injections or reduced production are procured at short notice in order to offset any mismatch between production and consumption at the time of delivery. We discuss the design of the ancillary service and real-time markets in Chapter 2, and in the context of congestion management in Chapter 4.

Finally, the system operator is typically responsible for planning development of the transmission network.

1.4   MARKET DESIGN

The definition of standard products is crucial to make electricity trading possible. Since the value of electricity varies continuously over time,

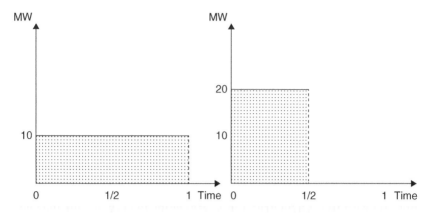

Figure 1.4   Consumption pattern of standardised hourly products

in the absence of any standardisation, the number of traded products would be unmanageably large: electricity delivered in one minute would be a different product with a different price from electricity delivered the following minute. Furthermore, each party would seek to trade different bundles of the many different products, because the pattern of each consumer's electricity withdrawal is different over time. As a result, transaction costs would be prohibitively high and inefficiencies would result.

Product standardisation means that, for the purpose of trading, different items are treated by all the market participants as identical products. In most wholesale electricity markets the basic standard traded product is the hourly-block. The hourly-block is the total production or consumption taking place over a fixed hour, irrespective of the pattern of production or consumption during the hour. The purchase of, say, 10 MWh in a given hour entitles the buyer to vary its consumption considerably during the hour. Figure 1.4 illustrates the effect of standardisation implemented on hourly products. The figure shows two of the many possible patterns of consumption of a buyer of 10 MWh in a given hour *t*. During the first half-hour the buyer could consume the entire amount purchased, that is, it could consume 20 MW for the first half-hour and nothing afterwards, as shown in the right panel. Alternatively, the buyer could consume the 10 MW at a constant rate throughout the hour, as shown in the left panel. In both cases the buyer is regarded as having consumed exactly what it bought on the market, that is, a total electricity volume of 10 MWh. The seller enjoys the same flexibility, as it can discharge its obligation to deliver 10 MWh, by implementing any production time pattern resulting in 10 MWh, of total injections during the hour.

Product standardisation in the locational dimension means that the seller can discharge its delivery obligations by producing the quantity sold on the market at any node of the network, and the buyer can consume the electricity purchased anywhere. Product standardisation is beneficial in as much as it increases market liquidity and lowers transaction costs. However, product standardisation may generate system operation costs. In our example, ignoring for simplicity's sake all other injections and withdrawals, we assume that the buyer consumes 10 MWh during the first half-hour, as shown in the right panel of Figure 1.4, while the generator delivers 10 MWh at a constant rate throughout the hour, as shown in the left panel. In order to maintain system security, the system operator must procure additional injections in the first half-hour, when the buyer's consumption exceeds the seller's production. Conversely, reduced injections will have to be procured in the second half-hour, when the seller's production is greater than the buyer's consumption. Locational standardisation also can generate system operation costs, in the event that the available network capacity does not allow the power generated at the delivery node selected by the seller to be transferred to the buyer's node. In this case the system operator has to reallocate production across locations in order to enable the parties to the transaction to exercise their right to deliver and collect electricity wherever they want.

In principle, product standardisation is sufficient to enable trading, as it reduces the number of products that market participants can exchange. However, in the electricity industry the number of standard products is still very large. In the case of hourly products, 24 different products are traded per day; furthermore, in some markets electricity delivered at each network location is traded as a different product. In addition, the expected conditions of demand and supply and hence the expected market-clearing price for a given time of delivery may vary dramatically at different times before delivery. This can be the result of unexpected generating unit outages, changes in the availability of renewable sources such as wind and solar power, and changes in the weather conditions affecting demand. Discovering via bilateral negotiations the market-clearing price of all the products traded and updating the assessment as new information becomes available involves significant transaction costs. Furthermore, imperfections in the price-discovery process could lead to major inefficiencies should the wrong set of generators be activated. Power exchanges where electricity transactions are centralised are the answer to this coordination issue.

Each end-customer's consumption is also assessed in terms of standard products. The consumption of standard products by larger consumers is typically measured directly, since hourly or half-hourly meters are generally installed on their premises. Consumption of standard products by smaller consumers are typically assessed conventionally, by allocating part of the consumption that took place over the longer metering interval to each hour.

We discuss the special arrangements governing electricity transactions in Chapter 2.

1.5 ELECTRICITY TRANSACTIONS

Technology determines the arrangements through which electricity exchanged on the market is delivered by the sellers and collected by the buyers. For most commodities, a consumer whose supplier fails to deliver the contracted quantity at the agreed time simply does not consume (at that time). This does not happen with electricity, since it is technically impossible to inhibit the buyer's consumption in the event that the seller fails to produce the corresponding amount of electricity. However, since the power system must be kept continuously in balance, the system operator has to make up for the electricity not produced by the seller and consumed by the buyer. The seller will then pay the system operator for the quantity that it was unable to deliver at the contractual time. In this way commitments to produce and consume entered into on the market are enforced financially.

In practice, each market participant's commitments to produce or consume the electricity respectively sold and bought on the market are separately enforced. Each party to an electricity transaction notifies the system operator the volume that it committed to deliver (the seller) or to consume (the buyer). The notification creates two independent obligations: one between the seller and the system operator, committing the seller to produce the notified quantity, irrespective of the buyer's actual consumption; the other between the buyer and the system operator, committing the buyer to consume the notified quantity irrespective of the seller's actual production.

This process is illustrated in Chapter 2.

1.6 GENERATION CAPACITY ADEQUACY

The economic mechanism driving investments in electricity generation capacity is conceptually the same as the one operating in all other industries. Persistently high electricity and ancillary service prices attract capital to the industry when existing capacity is below the equilibrium level;

persistently low electricity and ancillary service prices discourage capital accumulation at times when installed capacity is above the equilibrium level.

However, widespread concerns exist that investments in generation capacity may not be sufficient and that specific policy measures are needed to ensure that installed capacity is enough to match demand at all times.

A number of features, to varying degrees specific to electricity, motivate these concerns. First, some elements of the market design, industry regulation or industry practices may cause generators' revenues to be systematically insufficient to attract the efficient level of investment. Second, imperfections in the administrative process that sets the market price for electricity in the event of scarcity may bias the incentives to invest in generation capacity. Third, capacity adequacy concerns are sometimes motivated by the specific risk structure of the generation business, such that small changes in demand or supply conditions can have a dramatic impact on generators' profitability. While the first two issues call for mechanisms that integrate the generators' income in order to attract an efficient level of investment, the third issue can be handled by coordinating the timing of investments in generation capacity in order to reduce the risk for investors. A more certain environment is expected to reduce the rate of return required by investors, to the ultimate benefit of consumers.

In Chapter 3 we discuss the rationale for the introduction of capacity support schemes and analyse alternative approaches followed in different markets.

1.7   THE ROLE OF THE NETWORK

Electricity is transported on a transmission network from the place where it is generated to the place where it is used. Electricity flows on the network according to the laws of physics, and it is impossible to force power to follow predefined routes between a production node and a consumption node. The flows of electricity on the network depend mainly on how total consumption and production are spread across the different nodes of the network. As a consequence, the primary and often only way to modify power flows across the network is to reallocate production among the different generating units.

Congestion occurs in the event that the network is not capable of hosting the power flows matching the electricity market transactions. When this happens there is a limited possibility of delivering power generated in low-price areas to consumers located in high-price areas. As a consequence the value of electricity at different locations is different.

Arrangements differ in the way they induce market participants to deviate from the volumes they would produce or consume if transmission capacity were unlimited. Two general approaches can be identified. The first limits the set of transactions that market participants can enter to those corresponding to power flows that the network can safely host. This is achieved by allocating and enforcing a set of feasible transmission rights, that is, rights to inject and withdraw power at different network locations. As a result, the wholesale electricity market clears at different prices in different locations in the event of congestion.

The second approach consists of compensating market participants for deviating from their desired level of production and consumption. In this approach, the electricity market ignores any transmission constraints, so that the market participants can freely select where the electricity exchanged will be produced and consumed. Subsequently, if the corresponding power flows violate one or more transmission constraints, generators and possibly consumers are paid to modify the level of production and consumption that they had scheduled at the different locations. This practice is known as 're-dispatch'.

In some markets, the transmission network is split into large market zones, areas where congestion is more likely to result. Trading within each zone is unconstrained and intra-zone congestion is dealt with via re-dispatch, whereas trading across zones is limited by enforcing a set of transmission rights.

In Chapter 4 we illustrate the impact of network congestion on the wholesale electricity market outcome and look at how congestion management is implemented.

1.8   COMPETITION POLICY IN THE ELECTRICITY INDUSTRY

Market power is a primary concern in wholesale electricity markets for two broad reasons. The first is that electricity is a primary commodity purchased by every household and business, and its price is extremely important for the economy.

The second is that the unique technical and economic characteristics of electricity make wholesale electricity markets particularly prone to market power. Since electricity demand does not respond to price in the short term and electricity is not storable, when the system is tight even relatively small generators may enjoy significant market power. This happens because,

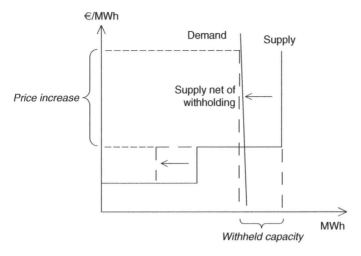

Figure 1.5   Generation capacity withholding when the system is tight

when existing generation capacity comes close to full utilisation, the with-holding of even a small quantity of supply from the market may cause a sharp increase in price. Figure 1.5 illustrates a situation in which the withholding of generation capacity leads to scarcity, that is, increases the market-clearing price to the level necessary to ration demand.[4] The major discontinuity between the system marginal cost (the market-clearing price in normal conditions), and the value of electricity for consumers (the market-clearing price in the event of scarcity) creates a strong incentive to exercise market power.

Transmission constraints add to the problem, since they reduce the scope for competition between generators connected in different locations. In addition, the owners of generating units in multiple locations may find it profitable to implement bidding strategies specifically aimed at creating network congestion, if this allows them to reap the benefits of lower com-petition at certain locations.

In some countries, market power in the wholesale electricity market is addressed by regulatory statutes. In the US, the federal regulator has the duty and the power to ensure that wholesale electricity prices are 'just and reasonable'. In Europe, charging excessive prices is addressed by competition law and considered an abuse of dominant position, although historically the prohibition on charging excessive prices has proved hard to enforce in this context.

In Chapter 5, we discuss how the specific technical features of electricity supply impact on the competition policy tools used to determine market boundaries and to assess the degree of competition in the industry. We also analyse the main types of policy measures implemented in electricity markets to mitigate market power, including asset divestitures, the imposition of long-term sales on the dominant generator, price caps and bid-mitigation mechanisms.

1.9   ELECTRICITY RETAIL COMPETITION

Like retailers in other sectors, electricity retailers bundle the inputs necessary to provide their electricity services to clients. Specifically, electricity retailers:

- are responsible for procuring from the wholesale market the electricity consumed by their clients;
- buy system operation, transmission, distribution and metering services;
- design and advertise offers addressed to consumers;
- act as an interface for their clients on matters related to the electricity service; and
- issue invoices and collect payments.

Since retail costs account for only a small share of the total cost of the electricity service, the potential for competition to reduce consumers' bills by reducing retail costs is limited. Therefore most of the benefits of retail liberalisation are linked to the ability of competitive retailers to provide services that are tailored to consumer preferences. Large consumers have diverse needs of energy price certainty, sophisticated procurement strategies and may possibly take advantage of their ability to control their use of electricity. They are therefore in a position to take full advantage of retail competition.

The benefits of a more diversified offer for smaller consumers are less evident, at least at this stage of the liberalisation process. In addition, smaller consumers appear to face significant transaction costs in order to identify, assess the offers of and switch to a different supplier. As a consequence, the incumbent retailer enjoys significant market power over its passive customers. This has led regulators in most jurisdictions to retain price controls long after the legal liberalisation of electricity retailing, and in some cases to question the opportuneness of retail liberalisation altogether.

We discuss electricity retailing in Chapter 6.

1.10 SUSTAINABILITY TARGETS AND THE FUTURE OF THE ELECTRICITY MARKETS

The main driver of the expected evolution of electricity systems, at least in industrialised countries and Europe especially, is the reduction of greenhouse gas emissions, which are responsible for the increase of global mean temperature.

In Europe, the Climate and Energy Package passed at the end of 2008 sets the 20–20–20 targets to be achieved by 2020: greenhouse gas emissions at least 20 per cent below 1990 levels, 20 per cent of energy consumption provided by renewable resources, and a 20 per cent reduction of primary energy use against a baseline scenario.

Given that the deployment of renewables in electricity is more cost efficient than in transport and – to a lesser extent – heating, the burden of the total renewable energy target placed on the electricity sector will be large. The production of electricity from renewable sources in Europe is expected to rise from 21 per cent in 2010 to 33 per cent in 2020.

The sustainability objectives appear to have radically changed the trade-offs relevant in the 1980s and in the 1990s when the decisions to liberalise generation activity were taken. At the time, regulated decision-making processes for deciding generation and transmission investments were perceived to be inefficient due to the incentives for regulated utilities to overinvest, errors in fuel price forecasts and political interference. These considerations and the reduction of economies of scale in generation, due especially to the development of combined-cycle gas turbine technology, were among the main motivations for the transfer of investment decisions from the government to the private sector.

However, the current policies addressing climate change increase political involvement in generation investment decisions. The development of renewable generation capacity is taking place under a variety of support schemes so that the size, composition and in some cases the location of new capacity are determined by public authorities. Most of the risk in investing in renewable generation capacity is placed on electricity consumers through measures that make renewable generators' revenues independent of market prices for electricity. The same holds in some countries for investment in nuclear generation, the viability of which appears to be increasingly dependent on the possibility of transferring part of the risk to electricity consumers via regulation.

It remains to be seen whether a hybrid system, in which planning governs the development of renewable and possibly nuclear capacity while conventional capacity is supplied by the market, is sustainable. The profitability of non-renewable capacity may be dramatically impacted by the level of renewable capacity set by public decision makers. In the event that such regulatory risk makes private investment in conventional generation capacity unattractive, electricity consumers could end up bearing the risk of all investment in generation capacity, for example in the form of more expensive capacity support schemes.

Renewable energy sources are intermittent, and their availability can be accurately predicted only a short time in advance of the time of delivery. As a consequence, the production programmes of an increasing share of generation capacity may have to be modified close to real time, as the expectations of renewable production are updated. This means that an increasing number of financial transactions need to take place close to the time of delivery, at prices that may significantly depart from those clearing the day ahead market. Moving generation scheduling decisions closer to real time might be particularly difficult in Europe, where electricity trading between market participants and the procurement of ancillary and balancing services by the system operator are carried out close to real time in separate venues with different rules. In this context, the consistency of the clearing prices of multiple markets run separately in a very short timeframe might not be achieved. This could result in an inefficient use of generation capacity.

Furthermore, if the trends currently observed continue in the future, network congestion will become a recurring feature and the need for fast-response generation capacity will increase. A greater need for re-dispatch to relieve congestion leads to an increase in total supply costs. This could lead to greater reliance on regulatory and administrative measures aimed at limiting system operation costs, with distortive effects on the prices prevailing in all markets cleared near real time. We explore these issues in Chapter 7.

The expansion of renewable production appears to be modifying the relative merits of the market and regulation in electricity generation. The assumption that the market is the most efficient way to govern the development of generation capacity is one of the drivers of the wave of liberalisation that took place in the 1990s and early 2000s. However, liberalisation can hardly be claimed to yield benefits in terms of short-term efficiency. On the contrary, a vertically integrated monopoly with unified control of the generation fleet and the transmission network makes implementing minimum-cost dispatch easier, compared with a market setting where the decisions of multiple independent generators are coordinated through price signals.

In this perspective, the politicising of capacity development decisions and exacerbating of short-term coordination issues brought about by the expansion of renewable generation reduces the market's relative merits compared with a centralised model.

It remains to be seen if competition can still play a significant role in the new environment, which is largely based on planning. Auctions could be implemented to select efficient renewable investments, for example. However, the main feature of the liberalised model, which is the allocation of investment risk to market investors rather than consumers, would be lost.

Finally, electricity demand is expected to contribute to the achievement of sustainability objectives. Policy measures to reduce consumption are being implemented in most countries, and further benefits are expected from an increase in the responsiveness of demand to spot prices. Many aspects of the technical, commercial and organisational arrangements enabling small consumers to respond to prices are still undetermined. The evidence that evolution of the electricity systems in that direction will deliver a net positive benefit is still weak. In any case, exploiting the full price-response potential of small electricity consumers will require massive investment and take a long time.

NOTES

1. KWh, MWh, GWh and TWh are commonly used multiples of Wh. KW, MW and GW are commonly used multiples of W.
2. Hydro peakers have a high opportunity cost, since their production is limited by the size of the reservoirs and needs to be allocated in time in the most valuable way. The opportunity cost of a hydro peaker is the cost that the system would bear if that capacity were not available. It can be thought of as the cost of the most expensive thermal generator that has been displaced by hydro production.
3. Typically the withdrawal nodes of a transmission network are the points of connection with distribution networks. Large industrial customers are sometimes directly connected to the transmission network.
4. A condition that we shall analyse in Chapter 2, Section 2.2.1.

2. Wholesale electricity markets

Guido Cervigni and Dmitri Perekhodtsev

2.1 INTRODUCTION

Trading electricity in a similar way to other commodities requires special arrangements. Transaction costs are reduced by product standardisation, and a central agent, the system operator, ensures that production and consumption match on a constant basis regardless of actions by generators and consumers. Figure 2.1 illustrates the typical timeline of electricity transactions.

In this chapter we discuss transactions taking place in wholesale electricity markets, splitting them into two groups: electricity market transactions and transactions related to system operations.

2.1.1 Electricity Market Transactions

This group includes transactions taking place between market participants up until shortly before the actual time production and consumption take place, which is referred to as the 'real time'. We refer to the broad range of such transactions as 'market transactions'.

All the market transactions are forward transactions, because they take place before the time of delivery. They are concluded between years

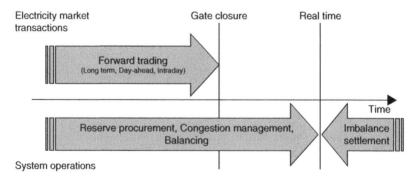

Figure 2.1   Timeline of electricity transactions

or months and a few hours ahead of real time. Products exchanged in market transactions are often highly standardised, both in terms of time and space. For example, a typical traded product is a volume of electricity produced and consumed in a given hour at any location in a given country. Production or consumption taking place at different time intervals within the hour or at a different location within the same national network are therefore considered identical products.

The arrangements governing forward transactions taking place near the time of delivery, generally starting from those taking place the day before delivery, are specifically designed to address the high volatility of electricity demand and supply. The price of electricity can differ dramatically from one hour to the next depending on the level of demand and on available generation capacity. In addition, the expected demand and supply conditions at a certain time of delivery may also change dramatically as the time of delivery approaches. In this context, decentralised or bilateral trading would entail high transaction costs and could lead to large-scale inefficiency. Organised markets, or power exchanges, reduce transaction costs and ensure an efficient market outcome by centralising the negotiations.

The latest time when market transactions for delivery at a certain time can be entered is called 'gate closure'. After gate closure the market participants inform the system operator which generating units they intend to activate in order to produce the electricity sold on the market and where in the network the electricity bought on the market will be consumed.

2.1.2 System Operations

The system operator concludes transactions with market participants as part of the activities it carries out to ensure that all the system's security constraints are constantly met.

Transactions related to system operations take place at different times around the moment of delivery. Starting from gate closure and through the time of delivery, the system operator procures additional production or reduced production on the real-time or balancing market. In the balancing market the system operator sells and buys energy from generating units and consumers with load-control capability, to match instantaneous imbalances between production and consumption. In order to ensure that there is sufficiently flexible capacity to perform balancing in real time, the system operator may procure reserve capacity in advance of the gate closure. Finally, both before and after gate closure, the system operator carries out actions intended to ensure that production and consumption

can be securely supported by the transmission network capacity. We refer
to these actions as 'system re-dispatch'.

The system imbalance, offset by the system operator in real time, is
the sum of the individual imbalances of market participants, that is, of
the deviations of market participants' actual production and consumption from the quantities they have notified to the system operator as
respectively sold and purchased on the market. Once the actual consumption and production of each market participant is known, based
on metering data, the system operator financially settles the imbalance
of each market participant. The prices at which imbalances are settled
are known as 'imbalance charges' and are closely related to the prices
prevailing on the balancing market. The process that determines the
price and the volumes of those transactions is known as an 'imbalance
settlement'.

All the transactions analysed in this chapter are highly interdependent.
Most of them refer to the same product: energy produced and consumed
at a given time. This is the case for the forward transactions between
market participants, the real-time transactions between the system operator and the balancing energy suppliers, and the transactions that settle the
market participant imbalances. In addition, some transactions involve
products that are jointly produced; in particular a generator's capacity can
be used either to produce electricity or to provide reserves.

The rest of the chapter is divided as follows. In Section 2.2 we discuss
market transactions. In Section 2.3 we address system operation-related
transactions. In Section 2.4 we compare the philosophies underlying the
electricity and ancillary service trading arrangements implemented on the
European and US markets.

2.2 FORWARD TRADING

Contrary to most other goods, even small changes in demand and supply
conditions have a major impact on electricity prices, since stockpiles
cannot be used as a buffer. Market participants hedge against price
volatility through forward trading. Electricity in the wholesale market
is bought and sold between power generators producing electricity from
power plants, retail suppliers serving end customers and traders that do
not control generating units or serve consumers.

Electricity is traded in the forward market in different timeframes,
ranging from some years to a few hours before the time of delivery. The
latest moment in time when market transactions for delivery at a certain
time can be entered is called 'gate closure'; typically the gate closure is set

Figure 2.2   Timeline of electricity transactions

one or two hours before real time. Figure 2.2 illustrates the portion of the electricity transaction timeline presented in this section.

In Section 2.2.1 we analyse the arrangements governing transactions entered into the day before delivery, the day-ahead market. In Section 2.2.2 we address transactions entered into on the day of delivery, the intraday market. In Section 2.2.3 we discuss longer-term transactions. Finally, in Section 2.2.4 we illustrate how electricity transactions translate into production and consumption commitments.

2.2.1   Day-ahead Markets

We begin our discussion of electricity forward transactions from the arrangements for day-ahead trading, because they are affected more than longer-term transactions by the technical features of electricity. On the day before delivery, market participants obtain relatively reliable information about the demand and supply conditions. In addition, some thermal generating units are rather slow to start producing power and the decisions to start up, or commit, these units need to be taken in the day-ahead timeframe. The design of the day-ahead market is crucial to the system's efficiency, since its outcome determines which generators will be started up and will therefore be available to match load in real time.

Demand and supply conditions may and do change after the day-ahead market is cleared. Producers and retail suppliers can adjust to such changes by trading in the intraday markets a few hours before real time. Increased electricity generation from intermittent renewable sources the availability of which is best predicted close to real time increases the importance of intraday markets.

This section is organised as follows. First, we discuss the rationale for centralising day-ahead transactions in power exchanges or pools. Second, we discuss the merits of the much-debated non-discriminatory or marginal price auction design implemented in most day-ahead markets. Third, we discuss pricing in conditions of scarcity. Fourth, we analyse how alternative market designs address the trade-off between the need for product standardisation on the one hand, and the need for generators to sell products that it is feasible to deliver given the technical constraints on the other.

Organisation of the day-ahead market: bilateral transactions, exchanges and pools

Much of the debate surrounding the restructuring of the electricity industry in the 1990s in both the US and Europe focused on how the electricity day-ahead market should be organised.

Three organisation schemes have received the most attention: bilateral markets, exchanges and pools. In a bilateral electricity market, buyers and sellers trade directly with no coordination by a central body. As a consequence, price discovery entails repeated one-to-one interactions between market participants.

An exchange is an entity that coordinates the trading of standard products between market participants. It collects buy bids (or simply bids) and sell bids (or offers) for electricity, and clears the market. Coordination provided by the power exchanges is crucial for efficiency in the electricity industry, because the value of electricity differs significantly from one hour to the next. Most exchanges trade hourly energy products with delivery on each hour of the day, that is, a total of 24 hourly products per day.[1] In the event that network-related constraints are enforced on market transactions, as we discuss later in Chapter 4, the value of electricity differs not only in time but also by location, and an even larger number of prices may need to be discovered by the market participants. For those reasons, discovering the value of electricity each hour via bilateral negotiations entails significant transaction costs. Exchanges reduce transaction costs by identifying the market-clearing prices and the corresponding series of transactions. However, trading outside the exchanges through bilateral negotiations is generally also allowed. Moreover, in some countries such as the UK, multiple exchanges clear the same physical market.[2]

When trading bilaterally or through an exchange, sellers do not make commitments as to which generating units will actually produce the electricity they have sold, and buyers do not make commitments as to where they will consume the electricity they have purchased. In contrast, pool market-clearing systems also perform scheduling. They determine the set of generation units that will be started up, as well as the production level of each unit in each hour and the hourly consumption by each market participant in each location.

Furthermore, the pool market-clearing algorithm ensures that the market outcome is feasible and secure. Feasibility means that the market-clearing production schedule for each unit is consistent with the unit's technical capabilities. In this respect, compared with an exchange, a pool provides a higher degree of coordination of market participants' decisions. Security means that the set of market-clearing schedules (i) satisfies all the network's security constraints, and (ii) leaves an adequate reserve margin for real-time balancing.

Those features have several major implications for the design of pool-based systems, ranging from the bid format to the market-clearing algorithm, which we investigate in the following sections and in Chapter 4.

With the partial exceptions of Italy and Spain, the day-ahead trading model currently prevailing in Europe is based on bilateral trades and trades through power exchanges. Market clearing and scheduling are carried out separately. After the gate closure, each seller decides which units will produce the electricity it has sold and each buyer decides where the electricity it has purchased will be consumed. These decisions are notified to the system operator in the form of injection and withdrawal programmes. After the schedules have been notified, the system operator takes the measures necessary to ensure that the system is secure.

In the rest of the chapter we base our presentation of the wholesale electricity markets on the European model, although in Section 2.4 we compare this model with the pool-based model implemented in much of the United States.

Market participants generally enter the day-ahead stage with open trading positions taken in the forward markets. They use the day-ahead market to adjust their forward positions. However, in order to simplify the presentation of day ahead markets we have assumed that no positions have been taken by the market participants prior to the day ahead, and we have ignored speculative purchases and sales. This implies that supply in the day-ahead market corresponds with the generators' production capacity, and that demand corresponds with expected consumption. Finally, we have assumed that bids and offers in the day-ahead market do not reflect the expected outcomes of the subsequent market sessions. In other words, we have assumed that no arbitrage takes place between the day-ahead and the intraday and balancing markets. We remove these simplifying assumptions in the following sections and chapters.

The auction model: non-discriminatory versus pay-as-bid auctions
In the day-ahead trading session, market participants submit offers for production and bids for consumption of electricity on the following day. The market operator accepts bids and offers in order to maximise the net gains from trade, or the surplus generated by the transactions.

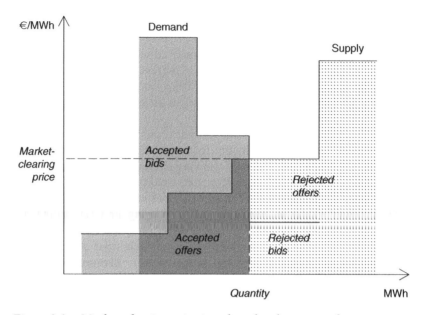

Figure 2.3   Market-clearing price in a day-ahead power exchange

For the sake of simplicity of exposition we have discussed alternative auction models in a market with only simple bids and offers. Each simple bid or offer refers to consumption or production in a single given hour.[3] In a market with simple bids and offers, the set of accepted bids and offers cleared for each hour is independent from the sets in the other hours. In this case the market equilibrium for one hour can be characterised as the point where the bid-based supply and demand intersect,[4] as shown in Figure 2.3.

The pricing and clearing rule implemented in the day-ahead markets has been the subject of extensive discussion.[5] All day-ahead power exchanges and pools that we are aware of run non-discriminatory auctions (also known as 'single clearing price' or 'single-price' auctions). In such auctions, every accepted bid and offer, respectively, pays and receives the market-clearing price, independently of the bid and offer prices.

Standard economic theory provides the basis for applying the market-clearing price to all accepted bids and offers. Provided that the market is competitive, in a non-discriminatory auction a generator's offer price reflects its incremental or marginal costs, while the bid price reflects the value of electricity to the consumer. The market-clearing price implements the set of transactions that maximises the gains from the trade and ensures efficient dispatch, that is, minimises total generation costs. The expected

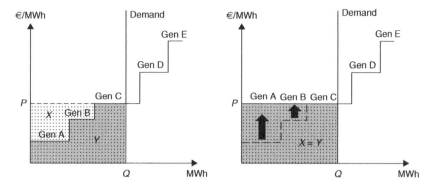

Figure 2.4   Bidding in single-price and pay-as-bid auctions

future clearing prices will drive efficient investment decisions, that is, cost-minimising generation technologies will be selected and new capacity will be built at the right time.

The alternative auction model is known as 'discriminatory' or 'pay-as-bid'. In a pay-as-bid auction, each accepted offer receives and each accepted bid pays their asking price.[6] For the sake of simplicity, we refer here to an implementation of the discriminatory pricing rule where buyers pay a uniform price equivalent to the average price paid to sellers, so that the market operator's budget is in balance.

Opponents to the single-price model often put forward a fairness argument. They argue that the single-price model results in excessive payments to generators, since low-cost units, such as nuclear, hydroelectric or even coal-fired power plants receive the same price as high-cost marginal units. This argument is flawed for several reasons. First, it relies on the assumption that generators submit the same offers in single-price and pay-as-bid auctions. In fact, rational generators bid differently in different auction models. This is illustrated in Figure 2.4. The left panel of the figure shows the competitive market outcome of the single-price auction. The generators' offer prices are equivalent to the corresponding variable costs and the clearing price is the price of the highest accepted offer. In this case, revenues collected by the generators on the market are represented by the sum of the areas X and Y.

According to the naive opposition argument, if the market cleared according to the pay-as-bid model, the generators would not change their bidding strategy. In this case their total revenues would fall to the area Y in the left panel in Figure 2.4. In fact, in a pay-as-bid auction rational generators would guess the price paid to the highest accepted offer – the clearing price in the single-price auction. They would then align their offers to that expected price. As shown in the right panel of the figure, there is no reason why generators A and B might want to sell at a price lower than *P*, since other generators, whose variable cost is higher than *P*, provide no competitive pressure to do so. If generators can correctly predict the clearing price, the total payments they receive will be approximately the same in both the single-price and the pay-as-bid auctions.

Note, incidentally, that convergence of offer prices to the market-clearing level is not restricted to pay-as-bid auctions. In ordinary markets where transactions are agreed upon on a bilateral basis, all transactions usually take place at prices that approximate the market-clearing price, despite variations in production cost among suppliers and differences in the availability to pay among consumers.

The second flaw of the argument in favour of the pay-as-bid model is that, even assuming that payments to generators are smaller compared with single-price auctions, if the market is sufficiently competitive such a reduction of the generators' revenues is not desirable. Higher gross margins for low variable-cost generators are necessary in order to remunerate the higher fixed costs of units that use less expensive fuel, such as hydroelectric, nuclear, solar and wind power. Without those margins, capital would not be attracted to electricity generation capacity and the system would eventually experience shortages.

A similar argument applies to another (alleged) virtue of pay-as-bid auctions: lower price volatility across hours. Price volatility is physiological in the wholesale electricity market, because different generators have very different variable costs and demand is volatile. It is therefore efficient for the market-clearing price to vary considerably in time. Extremely high prices when the market is tight are necessary to attract investments in peaking units, which produce a limited number of hours per year, and to induce demand to voluntarily reduce when supply is scarce. Moreover, forward and future contracts allow transfer of the price volatility risk to the parties that can bear it at the lowest cost.

In the presence of market power a reduction in the payments to generators may be desirable. Theoretical analyses[7] find that prices are lower under pay-as-bid than under non-discriminatory auctions in simplified settings, although to a much lesser extent than the naive hope that would expect each generator to cash in only its variable cost. The limited available empirical evidence shows no significant differences in prices between pay-as-bid and non-discriminatory auctions. Some experimental studies find higher prices in pay-as-bid auctions compared with non-discriminatory auctions.

While its superiority is not proven, some features of pay-as-bid auctions make them less attractive than non-discriminatory auctions in the electricity spot markets. First, the generators' profit-maximising strategy under pay-as-bid requires prediction of the market-clearing price. This prediction will necessarily be imperfect, and various market participants will forecast different values for the market-clearing price. As a result, a generator with a lower offer and high marginal cost will sometimes be selected rather than a generator with a lower marginal cost, and total generation costs will not be minimised.

Second, in a non-discriminatory auction the profit-maximising strategy for a competitive generator entails offering a price equivalent to the variable cost.[8] This relatively simple competitive benchmark facilitates the detection of market power, as large deviations of offer prices from variable costs can be taken as a signal of the exercise of market power. In some electricity markets, market-power mitigation measures are built around this competitive benchmark.[9] Specifically, market-power mitigation is obtained by administratively replacing the generator's offer prices with the estimated variable costs when offers of these generators are likely to constitute market-power abuse. In some US markets such as Pennsylvania–New Jersey–Maryland (PJM), California (CAISO) and Texas (ERCOT), automatic price-mitigation schemes are employed when offers are deemed non-competitive.

In contrast, in a pay-as-bid auction the competitive offer price systematically deviates from the variable cost. This makes the competitive benchmark more difficult to identify and the assessment of the exercise of market power highly controversial. As a consequence, under a pay-as-bid system, market monitoring would become more complex, and many of the current market-power mitigation measures would be inapplicable.

Pricing in conditions of scarcity

As long as available generation capacity is lower than demand, the clearing price of the wholesale electricity market equals the variable cost of the most expensive generator that needs to be activated to meet load, or the system marginal cost[10]. The wholesale market-clearing price is subject to discontinuity when demand is greater than the available generation capacity, that is, in conditions of scarcity. In this case, the competitive market-clearing price is no longer equivalent to the marginal generation cost; instead, the clearing price is the one that rations demand. We refer later to the difference between the market-clearing price and the system marginal cost as the 'scarcity rent'.

The steeper the demand curve, the bigger the difference between the price in normal conditions and the price in conditions of scarcity. Figure 2.5 illustrates this feature, based on the assumption that demand is perfectly inelastic in the relevant price range.[11]

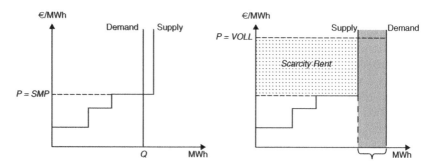

Figure 2.5   Scarcity rent

Currently, a large portion of electricity demand is indeed price insensitive in the day-ahead timeframe. This reflects the typical retail supply arrangements that charge the same price for the consumer's total consumption over long time periods, ranging from one to several months. Most of the electricity meters currently in place do not record consumers' hourly consumption, but only total withdrawal over a longer time period. This makes charging a different price for consumption in different hours impossible. The insensitivity of short-term demand to price also reflects consumers' preferences, as the value of electricity to consumers is generally much higher than the typical production cost.[12]

Should a price increase not reduce demand, quantity rationing must be implemented in the event of scarcity. Such scarcity may occur, for example, in the event that the day-ahead market cannot clear because the price-independent demand bid quantity is greater than the offered quantity. In this case, the system operator plans curtailment of service on different portions of the grid (rolling blackouts).[13] Since selective disconnection is technically infeasible, all the consumers connected to the same network branch will be disconnected at the same time.

When scarcity occurs and demand is totally price inflexible, the price for electricity is set to an administratively defined value, the value of lost load (or VoLL). VoLL is intended to be the price that makes consumers indifferent between consuming electricity at that price, and not consuming. VoLL is typically estimated several orders of magnitude greater than average electricity prices.

The values of VoLL vary between countries. For example, in the UK the VoLL used is €4.18/kWh, in Italy it is €10.8 and €21.6 per kWh, respectively, for residential and business customers, in Ireland €7.2/kWh, in Norway €0.96, €11.8 and €7.9 per kWh, respectively, for residential, commercial and industrial customers, and in Sweden €12.0, €8.8 and €7.9 per kWh, respectively, for urban, suburban and rural customers.[14]

Alternative spot-market designs differ in the way the clearing price is set in conditions of scarcity. One approach relies on market participants offering prices higher than their variable costs when conditions of scarcity are expected. When the system is known to be stretched, each generator calculates a certain probability that demand will be greater than the total available capacity. The generator's expected profit-maximising strategy in that situation entails offering part of its capacity at prices greater than the variable cost. By doing so, the generator takes into account the possibility that its offer will set the market-clearing price. In this situation the generator bears the risk that conditions of scarcity will actually not arise, and that its offer will be displaced by the competitors' offers. Productive inefficiencies may also arise if some offers above production cost turn out to be displaced by cheaper offers submitted by less-efficient generators. Finally, in this model the market-clearing price may turn out to be greater than the system variable cost, even if conditions of scarcity do not actually come about.

In the alternative approach, a scarcity-pricing mechanism is included in the market-clearing algorithm: the market-clearing algorithm automatically sets the prices for electricity and for the operating reserve at the VoLL when conditions of scarcity are detected based on the offers and bids submitted by market participants. In this situation, the generators' competitive bidding strategy is to offer its variable cost, irrespective of the expected demand and supply conditions, because in the event of scarcity the clearing algorithm itself will set the price at the VoLL. This approach is consistent with the broader objective of reducing the scope for inefficiencies caused by prediction errors. This requires the market participant's profit-maximising bidding strategy not to depend on its expectation of the clearing price.

In Europe this approach was followed in the pool system implemented in England and Wales between 1990 and 2001. Based on the available capacity offered in the pool and the expected demand, the system operator would assess the probability of not being able to serve the entire load the following day. Based on that probability and on the VoLL, the system operator would compute the expected value of the scarcity rent. This would then be added to the system marginal cost to determine the market-clearing price. In Nordpool, if the level of available capacity is such that the system operator must provide additional supply out of its capacity reserves, then the day-ahead price is increased to the price cap, and prices in the intraday and balancing markets must be as high or higher. Some US markets such as New York (NYISO), New England (ISO-NE), California

(CAISO) and Pennsylvania-New Jersey-Maryland (PJM) have or are currently developing scarcity pricing mechanisms.

Pricing in conditions of scarcity is a crucial element of the wholesale electricity market's design. Since the available generation capacity is far greater than demand in most hours, the competitive market-clearing price very rarely departs from the system marginal cost. Therefore the generating units with the highest variable costs rely on the extremely high prices prevailing during very few hours of scarcity to cover their fixed cost. If prices fail to reach the VoLL under conditions of scarcity, this may discourage investment in production capacity. That is a reason of particular concern in electricity, since the value of electricity for the consumers is generally much higher than the cost of generation. We address generation capacity adequacy in greater detail in Chapter 3.

Inter-temporal constraints

The cost function of most large thermal generating units reflects inter-temporal or dynamic constraints. Dynamic constraints take several forms, including (i) start-up cost: a cost borne every time the unit is brought into service; (ii) minimum technical output: a minimum level of production that the unit must deliver for technical or environmental reasons once it is brought into service; (iii) maximum ramp-rates: constraints that restrict the difference between the unit's production in one hour and the next; (iv) minimum up time: the minimum time that the unit needs to run once brought to service before it can be switched off; and (v) minimum down time: the minimum time that the unit needs to remain switched off before it can be brought into service again.

So far we have considered simple hourly bids and offers that are accepted or rejected independently for each hour. If only simple bids and offers are traded, the market outcome might result in production commitments that could not be accomplished by generating units subject to dynamic constraints.

This possibility is illustrated in Figure 2.6, where we show a hypothetical set of accepted offers in the day-ahead market for a generator that offered its entire capacity (10 MW) in each of five hours at a price equivalent to its variable cost (€50/MWh). The generating unit has a start-up time of three hours and, once activated, must produce at least 1 MW. The generator's offers were accepted in hours 1, 4 and 5 but not in hours 2 and 3. The corresponding production schedule is not feasible for the unit, since it involves zero production for two hours, while it takes three hours to bring the unit back in service after shutdown.

In this situation, the generator's production, in some hours, must depart from the quantities sold in the market. The generator selects the

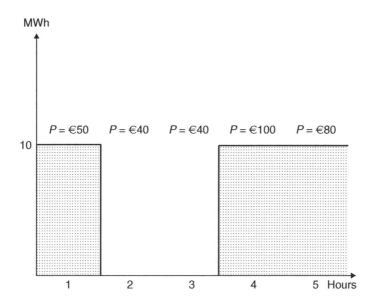

Figure 2.6   Infeasible market schedule

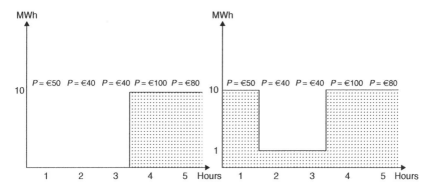

Figure 2.7   Feasible production schedule options

lowest-cost course of action between (a) starting up the unit only in hour 4 and buying replacement power to cover its cleared sales in hour 1, and (b) starting up the unit in hour 1, running it for all the hours at least at the minimum load, and selling the excess production in hours 2 and 3. The profit in both cases will depend on the price at which the generator is able to buy or sell replacement power. Figure 2.7 illustrates the generator production in the two cases.

A market where the products traded are not consistent with the genera-tors' cost function has two undesirable consequences. First, generators are exposed to risk. The profit-maximising offer strategy by a (competitive) generator depends not only on its cost but also on the expected market outcome. If the generator's forecast of the market outcome turns out to be incorrect, so that it has to produce at a loss or that it forgoes profitable sales,[15] the generator might wish it had offered differently. Higher risk for the generators, other things being equal, increases the expected rate of return necessary to attract investment in the generation industry, and as a result leads to higher wholesale electricity prices.

Second, the fact that each generator addresses its dynamic constraints independently of the others may result in production inefficiency. When multiple generators, each with its own expectation on the cost of adjust-ment, independently assess their cheapest adjustment strategy, the result-ing production decisions might not minimise system-wide generation costs.

Various solutions have been implemented in order to address inter-temporal constraints. The first is giving generators the opportunity to make adjustment trades after day-ahead market clearing. Generators can trade bilaterally after closure of the day-ahead markets or on the intraday exchange. Some markets, such as the one in Italy, provide special adjust-ment sessions run immediately after the day-ahead market outcome is known, where simple hourly products can be traded again. In these market sessions the generators can modify the positions resulting from the day-ahead market through additional purchases or sales, in order to commit to a feasible production schedule.

Second, most European day-ahead exchanges allow block order trading. A block offer (or bid) is a commitment to produce (consume) a constant amount of power in a group of consecutive hours at an average price no lower (greater) than the one specified in the offer (bid). Block bids and offers are either entirely accepted or entirely rejected. A generator sub-mitting, for example, a 12-hour block offer for its minimum technical capacity, at a price equivalent to the average generation cost including start-up cost, is sure that its unit will either operate above cost or not be activated at all. Because of the indivisibility of the block offers, the day-ahead market-clearing algorithm becomes more complex, as the impact of accepting a block bid or a block offer extends over multiple hours. Block products do not ensure that the market outcome is fully efficient or, from a different perspective, that no risk is put on the generators. Block products put the generator at risk of the chosen length of the block offer turning out not to be optimal. For example, a 12-hour block offer for a generator's minimum technical level at a price equivalent to its average cost (including start-up cost) might be rejected, whereas an eight-hour block offer at a price equivalent to the average cost might have been accepted.[16] In this case the generator offering the 12-hour block would forgo a profitable sale, and total generation costs would not be minimised.

The third approach is implemented in the pools, where generators submit offers that closely mimic their cost functions and the market-clearing algorithm explicitly takes into account the dynamic constraints of each generating unit. As a result, the market outcome is such that overall generation costs are minimised and each unit is allocated a technically feasible production programme.

2.2.2 Intraday Markets

Demand and supply conditions may change after the day-ahead market has cleared. Generating unit outages may reduce available capacity. As the time of delivery approaches, the availability of renewable sources such as wind and sunshine may change compared with the expectations on which day-ahead trading decisions were based. Electricity demand from industrial customers may vary according to the requirements of production processes, while demand from small businesses and residential customers may vary with changes in weather conditions.

Intraday markets allow market players to carry out transactions up to a few hours before real time. For example, a wind generator may become aware that a wind drop will make production of the volumes sold in the day-ahead market impossible. The generator can then honour its delivery commitment by purchasing the volumes it cannot produce on the intraday market.

Two broad models have been implemented in the intraday timeframe. The first relies on a non-discriminatory auction similar to the one clearing the day-ahead market. The clearing of the intraday market sessions occurs at regular intervals during the day of delivery. In Europe this approach is implemented in Italy and Spain. In the US standard market model non-discriminatory auctions clear the market every five or 15 minutes.

An alternative model implemented in the intraday timeframe is continuous trading. With continuous trading, market participants post offers and bids on an electronic billboard managed by the market operator. Each time a bid is submitted, its price is compared with the prices of the offers already posted and not yet matched. If one or more opportunities for positive net-value deals are available, the bid is matched with the offer that maximises the transaction's net value, that is, with the lowest-priced offer. The price for the transaction is set as the price of the offer picked from the billboard to clear the bid. If no opportunities for positive net-value deals are available, the bid submitted is left posted in the repository. The same happens when an offer is submitted.

Support for continuous trading in the intraday timeframe comes particularly from traders, who see the price volatility as an opportunity for profitable deals around real time. Auctions might instead reduce the value for a trader of being able to access and process the information about demand and supply changes more quickly than others.

The drawbacks of continuous trading are of the same type as those discussed in Section 2.2.1 with regard to decentralised trading: the sequential matching process and the pay-as-bid pricing rule do not ensure an efficient market outcome and price discovery. Furthermore, in the current implementations of continuous trading for cross-border transactions, network-related constraints are enforced by clearing transactions on a first come, first served basis: bids and offers are matched across countries based on the order of arrival on the billboard until the available cross-border capacity is fully utilised. As a consequence, the set of cross-border transactions resulting from continuous trading might not be the one that makes efficient use of the transmission capacity in the event of congestion.

The intraday market implemented in Nordic countries by Nordpool gives market participants the opportunity to continuously trade hourly power products, as well as block orders. Trading takes place every day until one hour before delivery. Intraday transactions in Nordpool account for around 1 per cent of all spot transactions. Organised intraday market based on continuous trading design is also applied in the Central Western European market combining France, Germany, Belgium and the Netherlands. Furthermore, bilateral trading can be carried out in the intraday timeframe.

2.2.3 Long-term Transactions

While short-term trading is meant to ensure that electricity production costs are minimised at all times, long-term transactions play a key role in sharing the risk among market participants.

Electricity generation is highly capital intensive and highly risky. For that reason the possibility for generators to transfer risk via long-term contracts is crucial in attracting capital to the industry at minimum cost. The level of long-term contracting is commonly regarded as an indicator of the maturity of a wholesale electricity market.

Long-term contracts for electricity are traded similarly to those for most other commodities. In Europe, long-term wholesale electricity contracts are traded in over-the-counter (OTC) markets (that is, directly between the counterparties), as well as through organised power exchanges where transactions are mainly financial and products are standardised.

The volumes of long-term contracts traded vary significantly between European countries. In 2010 the volumes of long-term contracts traded in Germany and France via power exchanges amounted to around 500 TWh.[17] In Italy only physical contracts are traded, and the volumes traded over the counter and via power exchange amounted to around 310 TWh in 2010. NASDAQ OMX Commodities Europe is the most developed European forward market, allowing trading of financial derivatives contracts on the Nordic, German, Dutch and UK power markets. In 2010 there were transactions for 3,400 TWh, of which 2,100 TWh were over the counter and 1,300 TWh via power exchange.

The range of products traded includes base- and peak-load futures, forwards, options and contracts for difference. The trading time of these products ranges from a week to six years ahead of time of delivery. On power exchanges most traded products are futures contracts and options settled by financial payments rather than physical delivery.[18] These transactions are highly standardised with respect to the contract specifications, trading locations, transaction requirements and settlement procedures.

2.2.4 Market Position and Physical Nomination

We have described above the sequence of transactions among market participants starting from long-term to day-ahead (D-A) and intraday (I-D) transactions. Trading in the different timeframes allows market participants to update their contract positions as the expected economic conditions at the time of delivery change.

At gate closure, trading between market participants stops. At that moment, the sum of all the contract positions of each participant with respect to the given delivery hour determines the physical obligation of that market participant in the delivery hour: a party with a net seller position has to produce a matching volume of energy from its generating resources; a party with a net buyer position is committed to consuming the matching volume of energy.[19] At gate closure each market participant must submit its physical nominations, also called 'production/consumption programmes', to the system operator, specifying which generating units will produce the electricity to match its net seller positions, or where in the network the consumption matching its net buyer positions will take place.

Figure 2.8 is a simplified illustration of the evolution of the contract position of a hypothetical market participant with respect to a delivery hour, and conversion of the net contract position into the physical nomination at gate closure. The generator has sold 100 MWh under a

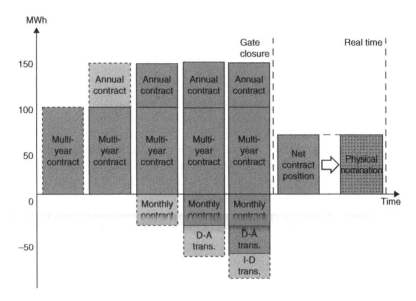

Figure 2.8   Market participant's net contract position at gate closure

multi-year contract and a further 50 MWh under an annual contract. A month before the delivery date it has the opportunity to buy 30 MWh at a price below its variable production cost. In the day-ahead market the generator buys a further 30 MWh at a price below its variable cost. On the delivery day an unexpected outage limits the generator's production capacity to 70 MWh. It then buys 20 MWh on the intraday market to make up for the reduced generation capacity. The generator's outstanding contract position at gate closure is given by the sum of all its previous positions and is equal to $100 + 50 - 30 - 30 - 20 = 70$ MWh. At this point the generator is required to match the 70 MWh contract position by notifying 70 MWh of production.[20] The balance of all the contract positions within a system is zero at any time, because each buying position has a matching selling position. Furthermore, since each market participant must notify a physical position that matches its final contract position, the balance of all the physical volumes notified at gate closure within a system is also zero, for each delivery hour. The following simple example provides an illustration of this.

Consider three market players. The first one is a utility that owns two generating units, Gen A and Gen B, and supplies two consumers, Cons 1 and Cons 2. The second one is a retail supplier that serves one consumer, Cons 3, but has no generation capacity. The third one is a trader that does not have any generation capacity or supply any consumers. Assume that

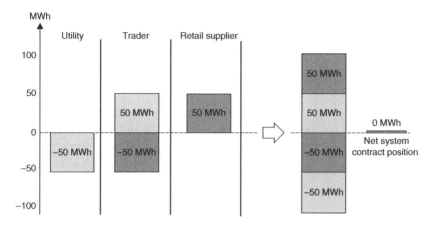

Figure 2.9   System net contract position at gate closure

the following set of transactions take place between the market participants. The utility sells 50 MWh to the trader; the latter sells 50 MWh to
the retail supplier. As a result of these two contracts, at gate closure the
utility has a net contract position of –50 MWh, the retail supplier of +50
MWh and the trader's position is zero. The combined contract position of
the system is zero, since the positions of the three players are matched.[21]
This is illustrated in Figure 2.9.

After gate closure, each market participant notifies the system operator
of its physical positions. The balance of each market participant's notifications must match the final contract position. The utility notifies 60 MWh
of production by Gen A, 40 MWh by Gen B, and 30 MWh consumption
by Cons 1 and 20 MWh by Cons 2. The balance of the utility's notification
is net production of 50 MWh, which offsets its final contractual position.
The retail supplier schedules 50 MWh consumption by Cons 3, matching
its contract balance, and the trader does not notify any physical position.[22]

The net physical position of each player matches its contract position.
Therefore, just as contract positions are balanced over the system, total
production and consumption over the system nominated at gate closure
are also balanced. This is illustrated in Figure 2.10.

The production and consumption programmes notified after gate
closure serve two important purposes. First, they represent each market
participant's commitment to deliver the electricity sold on the market and
to consume the electricity purchased on the market. In the next section we
discuss how those commitments are enforced.

Second, the notified production and consumption programmes tell
the system operator where in the network market participants intend to

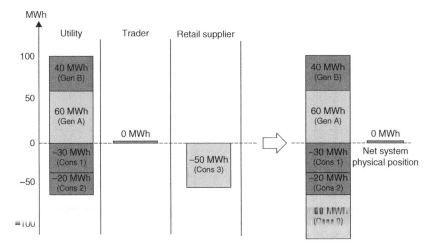

Figure 2.10   System net physical position at gate closure

produce and consume electricity at the time of delivery. This information is crucial to assessing whether the market participants' intended injections and withdrawals violate any system security constraints. In Section 2.3.4 and in Chapter 4 we discuss the remedial actions implemented by the system operator in such a case.

2.3   SYSTEM OPERATIONS

In this section we describe the activities performed by the system operator to ensure that production and consumption are in balance at all times. The three groups of system operation activities discussed in this section are illustrated on a time scale in Figure 2.11. They are imbalance settlement, balancing, and operating reserve procurement.

The objective of imbalance settlement is to identify any differences between each market participant's production (or consumption) commitments and actual production (or consumption), and to charge for them a price reflecting the value of electricity at the time of delivery. The objective of the balancing activity is to procure the energy needed to offset imbalances in real time. The objective of reserve procurement is to ensure that sufficient capacity will be available in real time in order to perform balancing. Both before and after gate closure the system operator carries out congestion management actions in order to ensure that that network's security constraints are met.

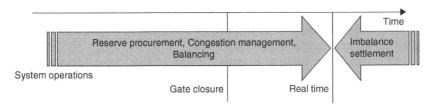

Figure 2.11   The timeline of energy delivery and ancillary services

We discuss system operations activities in reverse chronological order, starting from imbalance settlement (Section 2.3.1), moving on to balancing activities (Section 2.3.2), and finally the forward procurement of operating reserves before gate closure (Section 2.3.3). The discussion of the system operation activities in the first three sections ignores network security constraints. In Section 2.3.4 we discuss how system operation activities are impacted by network security constraints, and in Chapter 4 we analyse in greater detail the economic implications of network security constraints.

2.3.1   Imbalance Settlement

Production and consumption programmes notified after gate closure represent each market participant's commitment to deliver the electricity sold on the market and to consume the electricity purchased on the market. However, there is no guarantee that at the moment of delivery the exact nominated volumes will be produced and consumed and that there will be a perfect match between total production and total consumption in the system.

Departures of actual consumption and production from the notified volumes can and do happen after gate closure. For example, such imbalances could happen because of unforeseen generating unit outages, unexpected changes in weather conditions affecting units' performance,[23] and the availability of intermittent renewable energy sources such as wind and sunlight. On the consumers' side, imbalances reflect varying electricity requirements.

For most goods, a consumer whose supplier fails to deliver the contracted quantity can choose either to give up consumption of the missing quantity or to procure it from an alternative supplier. In the electricity markets, the latter option is implemented. However, because of the very short timing of real-time operations, such alternative procurement is centralised. In the event that the delivered or consumed quantities do not match the contracted quantities, the system operator is in charge of procuring the replacement energy or disposing of the excess energy. This activity entails a series of transactions in which the system operator is the counterparty to the market participants.

The imbalance settlement process consists of two steps: assessment of each market participant's imbalance volumes and setting of the prices charged for the imbalance volumes. Below we present the main elements of imbalance settlement systems. We conclude this section by presenting the arrangements, known as load profiling, conventionally implemented to determine the smaller consumers' hourly electricity consumption for the purpose of assessing their suppliers' position on the wholesale market.

Imbalance volumes

After energy is produced and consumed, the system operator assesses the energy imbalance volume for each market participant as the difference between its net nominated volume and its actual net energy production or consumption. Nominated volume is the net quantity of electricity that a market participant is committed to produce or consume as a consequence of its sales and purchases on the market, as notified to the system operator at gate closure. Actual production and consumption are metered at the generator and consumer premises.

A market participant runs a positive real-time imbalance if it produces a quantity of electricity greater than its net sales on the market, or if it consumes less than its net purchases. Through the imbalance settlement process, excess production is sold to the system operator. Likewise, a participant running a negative real-time imbalance buys replacement energy from the system operator.

We shall use the example from the end of the previous section to illustrate the imbalance volume assessment. We assume, in addition, that the actual production and consumption of market participants in a given delivery hour is different from that nominated at gate closure and therefore different from each participant's net contract position. The utility's clients Cons 1 and Cons 2, respectively, consume 30 MWh and 25 MWh, compared with the nominated quantities 30 MWh and 20 MWh; the production of Generator A is 70 MWh instead of the nominated 60 MWh, and the production of Generator B is 10 MWh instead of the nominated 40 MWh. Consumption by the retail supplier's Cons 3 is 45 MWh instead of the nominated 50 MWh.

Figure 2.12 illustrates the calculation of the utility's imbalances in our example as the difference between net actual production and net nominated volume. The actual physical balance for the utility given by the sum of the production of the utility's generating units and of the consumption of the utility's clients is 25 MWh. The net contract position and net

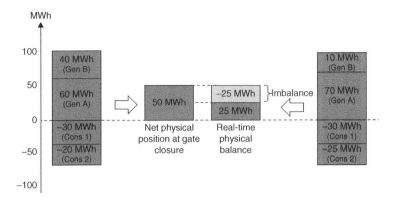

Figure 2.12   Imbalance of the vertical utility

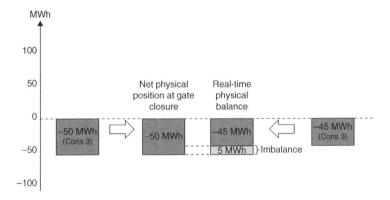

Figure 2.13   Imbalance of the retail supplier

nominated volume at gate closure was 50 MWh. The utility has therefore delivered 25 MWh less than it has sold in the market and notified. The system operator assesses the negative imbalance of 25 MWh, an amount that the system operator had to supply in real time in the utility's place.[24]

Figure 2.13 illustrates a similar calculation of the imbalances for the retailer in our example. The retailer's customer has consumed 5 MWh less than the amount purchased by the retail supplier at gate closure. The retailer is then assessed as having a positive 5 MWh imbalance.

Each market participant's imbalance volume is known only after the fact, when metering information becomes available. Before that, in real time the system operator has offset the total system imbalance, that is, the sum of actual production and consumption of all participants, by

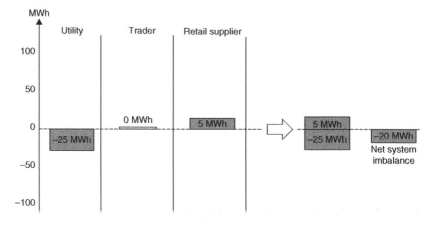

Figure 2.14   System-wide imbalance

Table 2.1   SO transactions in a simple imbalance settlement system

SO transactions	Volume, MWh	Price, €/MWh	Value, €
Purchase on the real-time market	−20	50	−1,000
Imbalance settlement: utility	25	50	1,250
Imbalance settlement: retail supplier	−5	50	−250
Total	0		0

procuring additional production or reduced production. As shown in Figure 2.14 for our example, the total system imbalance is the sum of all the market participants' individual imbalances.

Imbalance prices
The prices at which imbalances are cleared with the system operator are called 'imbalance prices'. Efficient imbalance prices reflect the value of electricity at the time of delivery. That value is the clearing price of the real-time market where the system operator buys and sells electricity in order to offset the total system imbalance.

Table 2.1 shows the outcome of transactions between the system operator and the market participants which run imbalances, and those carried out by the system operator to procure the energy to offset the total system imbalance. We have assumed that the real-time clearing price is €50/MWh The sum of energy volumes bought and sold by the system operator is zero because it buys the volume of energy on the real-time market that is needed to match the total system imbalance, that is, the sum of the market participants' individual imbalances. The system operator's budget is also balanced in our example, as all the transactions carried out to settle individual imbalances are priced identically.

In practice, assessing the value of electricity at the time of delivery for the purpose of setting imbalance prices is not as straightforward as in this example. One of the reasons is the temporal and spatial granularity mismatch between the imbalance prices and the prices of the transactions performed by the system operator on the real-time market. In the balancing market the system operator may have to buy or sell balancing energy at different prices in different intervals of the same hour[25] and at specific locations of the network,[26] whereas imbalance volumes are usually settled on an hourly or quarter-hourly basis, irrespective of where in the network the imbalance took place. Below we discuss how this mismatch is dealt with.

Dual imbalance pricing systems

Although trading between market participants stops at gate closure, they can buy or sell power after gate closure by running voluntary imbalances. For example, a generator may expect a large negative system imbalance in real time or, equally, that the clearing prices of the real-time market will be much higher than the day-ahead or intraday prices. It might then take the risk of producing more than the notified volumes. This would result in a positive imbalance by the generator that would be purchased by the system operator at a price reflecting real-time market prices, as illustrated in the previous section. Likewise, a generator expecting real-time prices lower than the day-ahead and intraday prices might decide to produce less than the notified volumes and to buy the energy deficit at the imbalance price.

In most European markets, with the exception of the Netherlands, such behaviour is discouraged through specific imbalance pricing arrangements. The main rationale for discouraging voluntary imbalances is the mismatch between the temporal and locational granularity of the prices in the real-time balancing market and the prices charged for the imbalances. Because of this mismatch, the price charged for an imbalance may not correctly reflect the cost, or reduction in cost, effectively caused by the imbalance in real time. This may therefore create distorted incentives to market participants to run voluntary imbalances, especially in the event of congestion in the network, as we discuss further in Section 2.3.4 and in Chapter 4.

Market participants are encouraged not to run voluntary imbalances by specific imbalance-pricing arrangements that make imbalances less profitable than sales and purchases on the market, irrespective of the cost or cost saving that these effectively bring about. One such mechanism sets a different price for imbalances of a given participant depending on the positive/negative system imbalance. In the industry jargon these arrangements

Table 2.2   Dual-price imbalance settlement system

		System imbalance	
		Negative $(P_{BM} > P_{DA})$	Positive $(P_{BM} < P_{DA})$
Market participant imbalance	Negative	P_{BM}	P_{DA}
	Positive	P_{DA}	P_{BM}

are commonly referred to as 'dual imbalance pricing systems'. A stylised example of this approach which is implemented in many European countries is shown in Table 2.2. The imbalance charges presented in this table are based on the combination of the price of the day-ahead market P_{DA} and the price on the real time balancing market P_{BM}.

When the system imbalance is negative, the system operator purchases additional power on the real-time balancing market to balance energy in the system. The real-time price on the balancing market in this case is likely to be higher than the day-ahead price. Market participants that run negative imbalances pay the balancing market price, which is higher than the price they would have paid on the day-ahead market to purchase the electricity that they did not deliver (or that they consumed in excess of their purchases). Market participants running positive imbalances and reducing the system imbalance receive the day-ahead market price and do not profit from their behaviour. They would have obtained the same profit by selling the additional volumes on the day-ahead market.

When the system's net imbalance is positive, the opposite applies: market participants responsible for positive imbalances receive a balancing market price that is lower than the day-ahead market price for production in excess of notification, while market participants running negative imbalances pay the day-ahead market price and thus do not benefit from helping to balance the positive system imbalance.

In this model the imbalance prices are explicitly designed to eliminate any possibility for market participants to profit from a voluntary imbalance in any direction.

Examples of such systems can be found in France, Belgium and Italy. The dual cash-out system that is used in the UK is similar. A participant with a negative imbalance, when the system imbalance is also negative, pays the System Buy Price, determined by the price of accepted offers on the real-time balancing market. This price is typically higher than the forward market price. A market participant with a positive imbalance, when the system imbalance is also positive, receives the System Sell Price based on the accepted bids on the real-time balancing market. This price is normally

Table 2.3   SO transactions in a dual-price imbalance settlement

SO transactions	Volume, MWh	Price, €/MWh	Value, €
Purchase on the balancing market	−20	50	−1,000
Charge to the vertical utility	25	50	1,250
Payment to the retail supplier	−5	30	−150
Total	0		100

lower than the forward market price.[27] Finally, when the participant's imbalance is the opposite of the system imbalance, the participant pays or is paid the reverse price, which is a market index price based on short-term energy trades on the within-day spot markets, like the day-ahead price in the system described above.

Consider the example above, assuming a dual-price imbalance settlement. In our example, the system imbalance is negative. We assume that the price in the balancing market is €50/MWh and the price in the day-ahead market is €30/MWh. Table 2.3 summarises the value of the transactions carried out by the system operator on the balancing market and in the imbalance settlement. In this case, the utility pays a high real-time imbalance price for the negative imbalance, while the supplier gets paid a lower day-ahead price for its positive imbalance. The table shows that in the case of the dual-price imbalance mechanism, the system operator budget no longer automatically balances and shows a surplus. Such surpluses are generally passed on to the network users, for example in the form of lower transmission tariffs.

A theoretical study carried out by Vandezande et al. (2010)[28] analyse the main impacts of dual pricing on wholesale trade. In addition to the lack of balance in the system operator's budget, the study mentions possible discrimination in favour of larger players in the event that their imbalance volume is assessed at portfolio level. Larger market players sustain lower imbalance costs because they have more opportunity to self-balance within their portfolio. The dual-price schemes provide incentives for inefficient strategies such as overcontracting in the wholesale market, withholding services for own use and nominating less than the expected production.

Portfolio aggregation for imbalance settlement
When different prices apply to negative and positive imbalances, market participants face different imbalance costs, depending on the degree of aggregation of production and consumption for the purpose of imbalance volume assessment.

In the example above, the imbalance volume was calculated at the

Table 2.4   Imbalance settlement of utility on a plant and customer level

Energy account	Notified volume, MWh	Actual volume, MWh	Imbalance volume, MWh	Price, €/MWh	Imbalance Cost, €
Generator A	60	70	10	30	300
Generator B	40	10	−30	50	−1,500
Customer 1	30	30	0	–	–
Customer 2	20	25	−5	50	−250
Total			−25		−1,450

market participant's portfolio level, that is, as the sum of the imbalances of all generating units and consumers under the market participant's responsibility. Under portfolio-based imbalance volume assessment, how production is split between Gen A or Gen B is irrelevant to the utility. That is because, given the consumption of the utility's customers, only the sum of the two units' production is relevant in assessing the imbalance volume.

However, reallocation of production between generating units in real time may create additional system operation costs, especially if it leads to violations of network security constraints. The system operator may then want to discourage voluntary imbalances, not only at the aggregate portfolio level, but also at the level of individual generating units and consumers. For this reason, in some markets the system operator assesses the imbalance volume at a finer level of granularity.

For example, the imbalance volume can be assessed separately for the portfolio of generators and for the portfolio of consumers under each market participant's responsibility. Alternatively, in some markets the imbalance volumes of large generating units are assessed individually, while the net imbalance is assessed for large sets of consumers.

Consider the vertical utility in our example above. At portfolio level it runs a negative imbalance of 25 MWh. In the case of dual imbalance prices in the example above, this imbalance will be charged at a price of €50/ MWh, and the total cost imbalance for the utility will be €1,250. However, the negative imbalance of 25 MWh is the sum of positive and negative imbalances at the level of individual plants and customers, determined as the difference between notified volumes and the volumes achieved in real time for each generator and customer. Table 2.4 summarises the imbalance settlement for the utility performed at the level of individual generators and customers.

Generator A creates a positive imbalance that is charged at a lower price of €30/MWh, whereas the negative imbalance induced by Generator B and Consumer 2 is charged at a higher imbalance price of €50/MWh. As a result, the total imbalance cost of the utility in this case is higher than if it were assessed at portfolio level. Even if there was a perfect energy balance within the utility's portfolio, the entity would still face imbalance costs if it distributed production between generating units differently from the way it nominated production at gate closure.

European markets that assess producers' and consumers' imbalances separately include Spain, Italy and the Nordic countries. In Italy, imbalances are assessed at the generating unit level for large generators and at plant level for large customers. On the contrary, for the purposes of assessing imbalances, injections by smaller units and withdrawals by small consumers are netted out. In the UK, individual generator energy accounts are calculated for each power plant. In France, Germany, the Netherlands, Austria and Belgium, imbalances are assessed on the total portfolio, which combines generators and consumers. In addition, multiple market participants can form a balance responsible party, a legal entity taking financial responsibility for the imbalances of consumers and generators within its perimeter.

Load profiling

Assessing the imbalance volume of a market participant involves comparing the notified volumes with the actual production and consumption volumes during each hour or smaller time interval. We have referred to this interval as the 'balancing period', and have assumed hourly balancing periods to simplify the description.

The actual output of most power plants is metered continuously. Therefore assessing the actual production of generators in each 'balancing period' presents few technical issues. Assessing consumer imbalances is less straightforward. Consumption by small electricity customers is rarely metered on an hourly basis. Conventional mechanical meters that are still widely in use can only record the total energy consumption between two meter readings that are typically taken on a monthly or multi-monthly basis.[29]

However, being able to assess the hourly consumption of each end-user is a necessary condition for implementing retail competition. It would be impossible otherwise to assess whether a volume of electricity matching the consumer's withdrawal had been procured on the wholesale market by the consumer's supplier. Since the value of electricity on the wholesale market is different in each balancing period, the withdrawal for which the retailer is responsible, that is, its clients' electricity consumption, must be assessed for each and every balancing period.

For this reason, arrangements referred to as 'load profiling' have been

Table 2.5   Load profiling of a non-hourly metered customer, MWh

	Past consumption	Hour 1	Hour 2	Hour 3	Hour 4
All non-hourly metered customers	5,000	1,000	1,200	1,400	800
Customer A	100				
Share of consumption allocated to Customer A	2%				
Hourly consumption allocated to Customer A		20	24	28	16

developed in order to calculate and settle consumption by non hourly metered consumers. In simple terms, for each hour each consumer is allocated a share of the withdrawals performed by the entire set of non-hourly metered customers. The consumer's share of the total consumption is determined on the basis of its consumption history. The total non-hourly meter consumption for each hour is calculated as the difference between total production and total hourly metered consumption.[30]

Table 2.5 illustrates this process in an example where, for reasons of simplicity, we have assumed that the total consumption of each non-hourly metered costumer is measured not every month, but every four hours. The table shows that the past metered consumption of Customer A represented 2 per cent of the past consumption of all non-hourly metered customers. Therefore the customer has been allocated 2 per cent of consumption by all non-hourly metered customers in each of the following four hours.

The retail supplier serving Customer A will be considered to have an imbalance equivalent, for each hour, to the difference between the nominated consumption and the consumption allocated to the consumer by load profiling.

At the end of the fourth hour the total consumption by Customer A in the four hours is known. The metered consumption will generally be different from the consumption allocated by the profiling system, as the latter was based on the consumer's past share of total non-hourly metered consumption. Table 2.6 illustrates the approximation error that may occur as a result of load profiling. In this example the total consumption over four hours approximated by load profiling for Customer A is 88 MWh, while the actual metered volume of consumption over those four hours was 120 MWh. The 32 MWh difference represents the volume that Customer A has consumed in addition to what its retail supplier has notified on its customer's behalf.

Table 2.6   Volume reallocation, MWh

	Hour 1	Hour 2	Hour 3	Hour 4	Estimated volume	Actual volume	Diff.
All non-real-time metered customers	1,000	1,200	1,400	800	4,400	4,400	0
Customer A	20	24	28	16	88	120	32

Note, however, that the total consumption allocated by the load-profiling scheme is, by construction, equivalent to the actual consumption of the entire group of non-hourly metered consumers. That means that some other non-hourly metered customers have consumed 32 MWh less than they were allocated by the profiling system. These differences are then settled among the retailers of the non-hourly metered consumers, by applying a volume-weighted average day-ahead price. In our example the retailer serving consumer A will pay for the 32 MWh consumed by its client (but not allocated to it by the profiling system), while some other retailers will be paid for 32 MWh allocated to but not consumed by their clients. This process of settling the errors resulting from load profiling between retail suppliers is sometimes referred to as 'volume reallocation'.

We note incidentally that load profiling is not a way to address the implications of the lack of demand price sensitivity discussed in Section 2.2.1. The cost to a retailer of serving a load-profiled consumer does not depend on how the consumer's actual consumption is spread across the hours. The supplier is financially responsible to the system operator only for the time profile conventionally assigned to the consumer and not for the actual pattern of the consumer's withdrawals, which remains unknown. As a consequence, the supplier does not benefit from its load-profiled consumer shifting consumption from high- to low-price hours.[31] The same holds for the consumer, whose supply price cannot be made contingent on the actual pattern of consumption, which is not measured, but only on the total consumption for the month.

2.3.2   Real-time Balancing

As discussed above, even if volumes bought and sold in all market transactions between participants are balanced, production may not perfectly match consumption at the time of delivery. In this section we describe how the system operator procures energy to balance production with consumption in real time to ensure that the power system is secure at all times. We present first the different services supplied by the generators to the system operator in real time. Then we discuss the market arrangements through which those services are procured.

Balancing services
Some balancing happens automatically with little direct involvement of the system operator. This type of balancing is called 'automatic control'. Automatic control is driven by changes in some system physical conditions caused by the energy imbalance. Automatic control acts fast, providing the time for the system operators to give instructions for manual corrections, after which injections by units providing automatic control are restored to their scheduled levels, ready to respond to a new imbalance.

In Europe a distinction is often made between two types of automatic control: primary and secondary control.

Primary control is an automatic reaction to frequency deviation of some generating units that are already producing. Imbalance between production and consumption causes the system's frequency to deviate from the nominal value (for example, 50 Hz in Europe). Frequency tends to fall in the event of negative imbalance and to increase when the imbalance is positive. Units providing primary control automatically increase production when system frequency falls and decrease production when frequency increases, offsetting the imbalance that has caused the frequency change. Primary control is normally activated within 30 seconds following the disturbance. A generator providing primary control is expected to perform deviations from its production programme of very short duration, around 15 minutes, after which primary control is replaced by other types of control.

A frequency deviation causes all units capable of providing primary control to respond, irrespective of where the imbalance between production and consumption has occurred in the interconnected network. As a consequence, the automatic activation of primary control may change the power flows between the country that experienced the imbalance and the neighbouring countries, resulting in the unintentional import and export of electricity.

Secondary control is a system that remotely controls specially equipped generating units located in a control area, in order to regulate their output up or down depending on the interchange flows between the control area and the neighbouring areas. For example, in response to a sudden deficit in production in a control area, primary control would increase production by all the generating units connected to the network within and beyond the control area. This would increase the import flows into the area where the production deficit occurred compared with the levels of

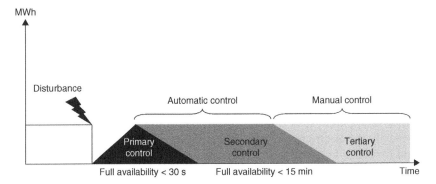

Figure 2.15   Sequence of automatic and manual control activation

cross-border transactions notified to the system operators by the market participants. Secondary control adjusts generation in the control area where the deficit has occurred until the cross-border flows are restored to their scheduled values. Generators supplying secondary reserve must be able to vary output within 15 minutes of the signal being sent and maintain the new production level for several minutes.

While automatic control is providing nearly instant response to the energy imbalance, the system operator gives instructions for manual changes to the production levels of the available units in order to offset the initial imbalance. In Europe, such manual control is often called 'tertiary control'. After manual corrections are activated, automatic primary and secondary control are no longer needed and are restored to their original levels in order to be ready for activation in the event of another contingency.

Figure 2.15 illustrates the typical sequence of activation of automatic and manual control in the event of a real-time imbalance, such as the outage of a large generating unit. The generator's outage causes a sudden drop in the system's frequency that triggers the activation of primary control. Within minutes the secondary reserve capacity located in the country where the outage has occurred increases output. As secondary control increases production, the frequency rises back towards the target level and the country's net imports return to their scheduled levels. Once the frequency reaches the target level, the production by the units providing primary control goes back to the target schedule and the primary control reserve margin is fully restored, ready for activation in case there is further need. Following the outage the system operator has also provided manual instructions to a slower generator to increase production. As soon as this tertiary energy control is implemented, production by the generators providing secondary reserve automatically reduces until their initial schedules have been fully restored. At that point production by the missing unit has been entirely replaced by the manually activated generators.

The balancing market

The energy produced by the primary reserve capacity is commonly not paid for, under the assumption that production increments and decrements, compared with the scheduled programmes, will balance out. Typically the electricity delivered by units providing secondary control is paid for either at a predetermined price or at the same price as the energy delivered by the generators providing tertiary reserve.

Manually controlled energy output increases and decreases are procured by the system operator on the balancing market, where generators, and possibly consumers, submit offers to provide upward and downward regulation.[32] The upward offers are prices that the generator agrees to receive for an increase in production from a specific unit compared with the energy schedule nominated at gate closure. Conversely, downward bids are the prices that the generator is willing to pay in order to decrease production from a specific unit compared with the nominated schedule.

If the system imbalance in the control area is negative, that is consumption exceeds scheduled production within the control area, the system operator eliminates the imbalance by accepting the necessary amount of cheapest upward regulating bids. Similarly, in the event of a positive overall imbalance in the control area, the system operator accepts downward regulating bids. Since accepting a downward regulating bid amounts to a sale of electricity to the bidder by the system operator, the latter selects the highest-priced downward regulating bids, in order to reduce the output of the most expensive units. This allows the system operator to keep the system balanced at minimum cost.[33]

As in the case of the day-ahead markets, there are two main options for settling accepted regulating bids and offers: the pay-as-bid and single-price clearing mechanisms. We looked at the relative merits of both approaches in the context of day-ahead markets in Section 2.2.1. The single clearing-price mechanism is generally regarded as more efficient for implementation in short-term electricity markets. However, for the balancing market many European countries have opted for the pay-as-bid system (for example, Italy, the UK, France and Belgium). The preference for the pay-as-bid system in the context of the balancing markets appears to be based on some specific features of the system operator's demand for balancing services, such as the need to accept bids and offers on a continuous basis and to address network security constraints, in addition to ensuring energy balancing, as described further in Section 2.3.4 and in Chapter 4.

The alternative design of the balancing market implemented in the United States is based on a series of non-discriminatory auctions, typically run every five minutes. Each session computes the system marginal price at each network location. That price is paid for all the offers accepted in that session and charged for all the bids accepted in that session. The same price is also applied to all imbalances. We discuss this approach in greater detail in Section 2.4.

Market arbitrage between the real-time and wholesale markets
Since electricity is non-storable, the price prevailing in the balancing market at each time is, strictly speaking, the only spot price for electricity. All transactions, with the exception of those in the balancing market, have a forward nature, given that electricity cannot be transferred from the seller to the buyer before real time, that is, when it is consumed.

Despite the fact that quantities traded in the balancing markets are generally small, the prevailing balancing prices, or real-time prices, may have a strong impact on prices in the wholesale electricity markets. Market participants can buy and sell electricity at real-time prices by making a bid or an offer on the real-time market, or by voluntarily running imbalances in real time and facing imbalance prices that are to some extent determined by the real-time prices (see Sections 2.3.1 and 2.3.4).[34] In other words, market participants have the opportunity to arbitrage between the wholesale and the real-time markets. No generator would want to sell on the wholesale market at a price lower than the expected real-time price, and no consumer would want to buy on the wholesale market at a price higher than the expected real-time price. As a consequence, any distortions in the real-time prices may filter through to the wholesale electricity prices.

2.3.3 Procurement of Operating Reserve Capacity

Real-time balancing services are provided by flexible generating resources that can quickly change output within the timeframe of the balancing market.[35] Flexible generators are, for example, hydropower plants and combustion turbines that can be brought into service at very short notice. Balancing services can also be provided by slow thermal generators that are already scheduled to operate below their maximum operating limit and can change their output within the required timeframe.

In order to ensure that in real time there is sufficient unloaded capacity to perform balancing, the system operators procure the reserve capacity long before gate closure. Generators providing such operating reserves undertake not to sell part of their capacity on the market and to keep part-loaded capacity available for the system operator's balancing purposes.

Below we discuss types of operating reserves as well as the cost of providing reserves for generators. We also discuss the existing approaches used by system operators to procure these reserves.

Types of operating reserves
Types of operating reserves differ in the dynamic characteristics of the reserved generating capacity. Generating capacity reserved to provide primary, secondary and tertiary control must be able to ramp up or down to the amount of reserve quantity provided, within the timeframe of the type of control that needs to be provided in real time, for example, 30 seconds in the case of primary reserves and 15 minutes in the case of secondary reserves.[36]

Therefore, the amount of each type of reserve that a given plant can provide is limited by its ramping rates. Because of the limit on the ramping rates, a single unit can rarely meet the entire reserve capacity requirement, and the system operator needs to procure reserve capacity from many units.

Resources supplying primary and secondary reserve capacity undertake to make the reserved capacity available in real time and to respond to the automatic control activation either upwards or downwards. Resources supplying tertiary reserve capacity are required to offer the corresponding production volume on the balancing market.

In most implementations the reserve capacity commitment does not fix the price for the electricity produced in the event of activation, that is, the price indicated in the balancing offers. However, the selection of reserve capacity offers may also take into account the price of electricity delivered in the event of activation. In this case, the price paid to generators to provide balancing energy, if activated in the real time, is also set by the reserve market[37]

Cost of providing reserve capacity
The cost of providing primary, secondary or tertiary operating reserves for generators is mainly determined by the opportunity cost of keeping capacity part loaded. In the case of reserve capacity for upward regulation, a generator needs to maintain the energy schedule below its maximum capacity, in other words providing 'capacity headroom'[38].

For example, if the spot-market price is €50/MWh, a generator with a variable cost of €30/MWh could receive a profit of €20/MWh by selling its power on the market. This means that providing 1 MW of upward operating reserve capacity per hour would entail opportunity cost of €20/MW per hour. This reserve provision opportunity cost arises when the wholesale market is profitable for the generator, that is, when the generator's variable cost is below the energy spot price. In the opposite case, when the energy market price is lower than plant variable costs, providing operating reserves may entail the cost of starting the unit up and the net cost of selling the minimum load at a price below the variable cost of production.

Opportunity costs are zero when the reserve capacity is provided by marginal and out-of-merit units, that is, by generating units with a variable production cost equivalent to or higher than the energy market-clearing price. However, the ramping rates limit the amount of reserve capacity provided by each individual unit. As a result, the requirement for reserve capacity cannot always be met by marginal units alone. Part of the reserve capacity is thus met by part-loading units with a high opportunity cost. For those units, the opportunity cost determines the cost of providing operating reserves.

Procurement of operating reserves
European system operators use two main ways to procure reserves. One is to buy reserves before the day-ahead market by acquiring unloaded capacity from market participants. After the reserves have been acquired it becomes the responsibility of the market participant to ensure that after clearing of the day-ahead and intraday markets the contracted reserve capacity is available. It is also up to the market participant to decide which units to use to schedule the reserve. Reserves can be contracted from years to hours before the time of delivery.

The other approach used by the Italian, Irish and Spanish system operators is to build the reserve margin after clearing of the day-ahead market when preliminary day-ahead plant schedules become known. After closure of the day-ahead market the system operators in these countries run special markets where they buy generators' deviations from the preliminary day-ahead schedules. For example, by purchasing a decrease of scheduled production by a unit, the system operator creates capacity headroom which will serve as reserve. These markets are called the 'restriction market' in Spain and the '*ex ante* MSD market' in Italy. They are also used to perform system re-dispatch before gate closure as we discuss below in Section 2.3.4.

The close interaction between the production of energy and provision of operating reserves provide the rationale for market arrangements for procuring these two products to be closely integrated. Integrated markets ensure the efficient allocation of generating capacity between the production of energy and operating reserves. If markets for operating reserves and energy markets are instead cleared independently, generators have to face risks, because assessing the profit-maximising offer price for one product means forecasting the price of the other. Forecast errors may then lead to inefficient offer and production decisions.

Perfect integration between energy and operating reserve markets is achieved on some of the US markets, where the spot energy market is cleared simultaneously with the reserves market. The system operator, which in the US markets also performs the function of market operator, selects the generators' offers for power production in order to optimally allocate generation to meet several objectives simultaneously: meeting the energy demand as well as the requirement for each type of reserve capacity. As a result of this simultaneous clearing, the system operator identifies the energy schedule and the operating reserves capacity schedule for each plant. It also identifies the market-clearing prices for energy and for each of the operating reserves.

In the European markets, the energy markets and the markets where the system operator procures operating reserves are typically cleared in sequence. The inefficiency of this approach is mitigated by bringing the market for operating reserves as close in time to the spot market as possible. This makes it easier for the market participants to arbitrate between the energy and the reserve markets, by increasing the quality of the predictions of the clearing prices of the two markets. This is done, for example, in Germany, where tertiary reserves are procured in a day-ahead timeframe, shortly before clearing of the day-ahead energy market (minute reserve). Generators submit offers to make capacity available in real time. The offers specify the quantity offered and the unit price. The system operator accepts the set of offers that meet the target levels of the different types of reserves at minimum cost. The approach implemented in Italy and Spain, where the system operator runs special markets shortly after the day-ahead market, also makes it possible to reduce the impact of the segregation of energy and reserve markets.

However, in many other European countries, operating reserves are procured by the system operator under long-term contracts. For example, in France primary and secondary operating reserves are procured under three-year contracts, and in Belgium, primary and tertiary reserves are contracted over four years, and secondary reserves over two years.

Other ancillary services
In addition to operating reserves, ancillary services also comprise the provision of reactive power and black-start capability.

The provision of reactive power is necessary to correct the voltage level on transmission lines. As the power flows from generators to consumers the voltage on the grid tends to drop. If the voltage drop is not corrected it may result in unsatisfactory operation of electrical equipment, causing damage to electrical motors. The voltage drop on a transmission line can be corrected by transformers that automatically adjust the voltage before

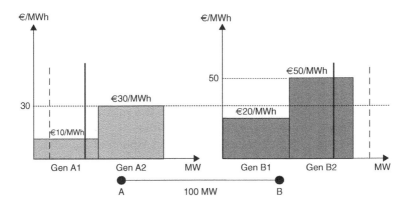

Figure 2.16   Example of re-dispatch

the power is distributed and through the injection of reactive power by generators. Reactive power is generally produced at a relatively low cost.

Black-start capability services are needed in the event of black-out to provide power to the grid in order to allow other plants to restart producing power. Most thermal generators cannot start up without taking electricity from the grid. Low-cost black-start capability is normally offered by hydropower plants.

2.3.4   Network Security Constraints and Market Design

Throughout the chapter we have mentioned that, in addition to matching production and consumption, the system operator guarantees the security of the network. When one or more network constraints are binding, the cost of meeting an increase in demand can be different at different nodes, as the example in Figure 2.16 shows. In this example, there are four power plants with capacity of 200 MW, each located in two interconnected zones. The variable costs of the two generators in zone A are, respectively, €10/MWh and €30/MWh, and the variable costs of the generators in zone B are €20/MWh and €50/MWh. Demand in zone A is 50 MWh and in zone B is 450 MWh. If there were no transmission capacity constraint between the two zones, the overall demand of 500 MW would be met at the least cost by generators A1, B1 and A2. If these generators bid their variable cost in the day-ahead market, the cost of generator A2 would set the day-ahead price at €30/MWh.

In this case, production in zone A would be 300 MW and production in zone B would be 200 MW. Given the consumption in each zone, this production plan would imply an export of 250 MW from zone A to zone

B. This export might not be feasible in the event that the network capacity between the two zones was 100 MW. Given this network security constraint, the least-cost option to meet the demand would be to produce 150 MWh in zone A and 350 MWh in zone B. In this case the marginal unit of energy produced in zone A would cost €10/MWh, that is, less than the clearing price that would be applied in the absence of the constraint, and the marginal cost of energy produced in zone B would be €50/MWh, that is, greater than the unconstrained clearing price.

In Chapter 4 we provide a more general description of the relationship between network constraints and the incremental cost of meeting demand at the different network nodes.

Alternative market designs differ in how they handle network constraints. The analyses carried out in this chapter thus far are consistent with an approach extensively implemented in Europe, in which the electricity market is run 'as if' no network constraints are binding[39]. In other words, market transactions are not subject to any network-related restrictions. This is achieved by standardising the traded product across locations. Market players buy and sell electricity knowing that the delivery and withdrawal obligations corresponding to their sales and purchases can be fulfilled at any location of the network of their choice. As a consequence, the energy market clears with a single system-wide price, as injections at any location are assumed to be perfect substitutes, as are withdrawals.

In this section we discuss the arrangements implemented within this model to ensure that system security conditions are met at all times. First we analyse how network congestion affects system operations, and then we go on to discuss how they affect imbalance price calculation.

System re-dispatch

When market transactions are not subject to any network-related restrictions, production and consumption schedules that market participants notify to the system operator at gate closure may be infeasible, that is, they may result in power flows that violate some transmission constraints. This situation is known as 'congestion'. Congestion is relieved by reallocating production between generators located at different nodes of the network, in order to ensure that the resulting power flows are within the network capacity limits.

This is achieved by paying generators[40] to modify their production from the levels notified at gate closure. These transactions result in different prices at different network nodes, because the congestion can only be relieved by increasing production at certain nodes and reducing production at certain other nodes.[41]

Such re-allocation is referred to as 'system re-dispatch'. System operators perform the system re-dispatch both after gate closure, during real-time balancing, and through various arrangements before gate closure.

In the real-time balancing market the system operator selects bids and offers to depart from the nominated volumes submitted by various plants in order to keep the system secure at minimum cost. In the previous section we assumed that the overall energy balance was the only relevant security constraint. In reality, in the real-time balancing market the system operator also seeks to meet the network security constraints. When one or more of such constraints appears to be violated, the system operator accepts bids and offers at selected network nodes until the constraint is relieved. In that case the system operator needs to pay location-dependent prices, acknowledging that production at different locations is not inter-changeable for the purpose of relieving a given set of network constraints. For example, a high-price offer may have to be accepted by the system operator simply because it is for a unit located in the same area that an imbalance occurred, while an identical imbalance at a different location could be offset by a much cheaper unit.

Locational pricing in the real-time market can be achieved in two ways: determining a location-specific clearing price that would be paid to or received from all accepted units in the same location, or clearing each accepted offer or bid at the offered or bid price, that is, using the pay-as-bid method. The former approach is implemented in the US markets, whereas the latter is used in most European balancing markets. As a result, the prices of the accepted balancing bids and offers may vary significantly across network locations.

Some network issues and constraints can be predicted by the system operator long before gate closure, for example, based on preliminary day-ahead production schedules obtained after the day-ahead market is cleared. It may be cheaper to address such congestion before real time by changing the commitments of cheap but slow units before gate closure, rather than performing re-dispatch of fast but expensive units in real time.

System re-dispatch applied before gate closure is used in several European markets, but approaches differ widely from country to country. Italy and Spain use a rather structured approach where, immediately after clearing of the day-ahead markets and after obtaining day-ahead generation schedules, the system operator runs an organised market for constraints resolution, accepting bids and offers from plants at specific locations in order to resolve network congestion. These markets are the *ex ante* MSD market in Italy and the restrictions market in Spain, and are the same markets where the system operator in those countries procures operating reserves.

Other countries use less-structured approaches. For example, in the UK the system operator may conclude a contract to acquire output from a particular plant before gate closure in order to resolve expected transmission issues. Such contracts are called 'Pre-Gate Closure Balancing Trades', or PGBTs.

The need for re-dispatch actions prior to market gate closure increases as congestion is expected to increase because of changes in the generating fleet, such as the deployment of a large amount of intermittent renewable power or phase-out of nuclear power not immediately matched by the necessary network upgrades. In several European countries, such as Germany, the Netherlands and Switzerland, system operators are currently considering introducing arrangements for constraint resolution before gate closure.

Early re-dispatch actions performed before gate closure may be followed by other trades between market participants in the intraday timeframe. Unless there are any specific restrictions, such trades may result in final nominations that restore the pre-re-dispatch flows, undoing the effect of early re-dispatch actions. Therefore, an effective early re-dispatch may require a mechanism enforcing the schedules achieved through re-dispatch throughout the intraday timeframe without unduly restricting the intraday trades. For example, one way to achieve this is to allow the system operator not to accept changes in final unit notifications compared with the preliminary day-ahead notification if such changes undo the effects of re-dispatch actions.

Imbalance prices in re-dispatch-based systems

In most European countries, market transactions are performed based on the assumption of unlimited transmission capacity, and are therefore not subject to any network-related restrictions. As a result, a single price clears electricity demand and supply in the entire country. Any re-dispatch cost incurred by the system operator in the event that market participants' intended production and consumption violate some network constraints is then socialised among all the network users.

The convention that location has no impact on the value of electricity needs to be carried over to the imbalance settlement stage, in order to avoid distortions in the market participants' behaviour and to limit the cost of balancing for the system operator. Consider, for example, the case in which imbalance prices are set at a national level in order to reflect the real-time balancing prices as described in Section 2.3.1 above. In the event of congestion, such imbalance prices would not be fully cost reflective. In the example in Figure 2.16 this means that the imbalance price charged by the system operator to generators for the negative imbalance in zone B would be lower than the price to increase production in this zone in order to resolve the imbalance. Likewise, the imbalance price paid by the system operator to generators for the positive imbalance in zone A would probably be higher than the balancing costs in this zone.

Imbalance prices set at a national level cannot therefore be relied upon to send correct signals to market participants of the cost caused by their imbalances in the event of congestion. In fact, making imbalance prices cost reflective would mean allowing them to vary by location. However, if energy markets are still cleared at uniform national prices, this would produce arbitrage incentives with potentially adverse wealth transfer from consumers to generators, as we discuss in detail in Chapter 4 Section 4.4. In addition, allowing arbitrage through voluntary imbalances could make system operations more complicated, because notifications would not provide reliable information about actual production at the time of delivery.

For that reason, in markets where congestion is dealt with via re-dispatch, imbalance pricing mechanisms like the dual pricing system discussed in Section 2.3.1 are implemented in order to discourage arbitrage via voluntary imbalances.

Once locational cost-reflective imbalance charges are ruled out, assessing the imbalance price entails some level of discretion. In the UK, for example, the system operator labels each transaction in the real-time market as either system or energy balancing. System-balancing transactions are those related to the resolution of network constraints, and are therefore removed from the calculation of imbalance prices.[42] Note, however, that the labelling system is highly conventional, since it is conceptually impossible to attribute a single purpose to real-time transactions that are chosen to address multiple simultaneous constraints.

2.4 WHOLESALE MARKET DESIGN IN EUROPE AND THE US

Throughout this chapter we have highlighted differences between the European and the US wholesale electricity market design. In this section we provide a brief analysis of the Standard Market Design, developed by the Federal Energy Regulatory Commission (the FERC) in 2003. Although this proposal was not officially implemented, many of the electricity markets currently operating in the United States follow the principles of the Standard Market Design.

The European and the US market designs differ in how they address the special technical features of electricity. The European design draws a clear line between the day-ahead and intraday energy markets on the one hand and the ancillary services and balancing markets on the other. The design of the day-ahead and of the intraday markets emphasises that electricity is a commodity; the special features of electricity are catered for by the ancillary services and real-time markets. In the US approach, on the contrary, the markets running from the day-ahead market through to real time are highly integrated. They share the same auction format, clearing algorithm and network security constraints. More importantly, they share the same product definition, which is fully consistent with the technical features of electricity. As a result, the specific technical features of electricity are consistently addressed in all market venues.

We compare next some features of the US and the European approaches as far as market transactions are concerned, to the extent that some common features can be identified in the very diverse arrangements implemented in Europe

2.4.1   Addressing Production Constraints

European energy markets are largely based on bilateral transactions and voluntary organised exchanges that trade standardised products, such as hourly bids and sometimes block bids. Central to US energy markets is a pool where generators submit the bids that are closely related to the individual generating plants, specifying the detailed plant dynamic characteristics in addition to the bid-based costs. The market-clearing algorithm takes into account the detailed constraints of the production fleet and looks for a solution that meets demand at the lowest cost while ensuring that production programmes are feasible.

2.4.2   Energy and Balancing Markets Integrated with Network Constraints

The US and the European approaches differ in how network security constraints impact the market outcome. The European energy markets impose no limits on trading within the entire market area, even in the event that such transactions result in power flows that violate network security constraints. Such network security issues are addressed by the system operators separately from market clearing through the sale and purchase of ancillary services or balancing actions.

In the US markets, congestion management is integrated with energy market clearing both in the day-ahead and real-time markets. The market-clearing mechanism provides a solution that meets demand at the lowest cost while satisfying all network flow constraints. In the event of congestion, the market-clearing prices differ across network locations (see further discussion on network congestion issues in Chapter 4).

2.4.3 Arbitrage between Day-ahead, Intraday and Real-Time Markets

In Europe, market participants are typically discouraged from taking speculative positions in the day-ahead or intraday markets, and close them on the real-time market. Such positions involve running voluntary imbalances in real time and paying imbalance prices. As we discussed in Section 2.3.1, imbalance prices often feature the dual-price system that makes such arbitrage unprofitable regardless of the market outcome. As a result, in the European approach arbitrage between the forward and real-time prices is prevented or considerably limited.

In the US, however, the real-time market and imbalance settlement are very tightly related. Imbalances are settled at real-time prices regardless of the direction of the imbalance. Unlike the European system, imbalance prices are locational and reflect the real-time cost of network congestion. Thus the incentives to run imbalances are properly aligned both on a system and on a locational level.

Furthermore, the US markets allow purely speculative purchases and sales in the day-ahead market that are settled at real-time prices. These speculative positions can be achieved through virtual bids submitted in the day-ahead market, that is, bids that are not associated with physical generating or consumption assets. Such arbitrage between the day-ahead and real-time markets results in day-ahead prices converging in the long term with real-time prices. As a result, day-ahead prices more accurately reflect the real-time market situation.

2.4.4 Integration of the Energy and Operating Reserve Markets

In the European model, energy is traded and reserve capacity is procured in separate market venues, running one after the other. As discussed above, this solution places the risk on the generators, since assessing the most profitable offer price for one product requires forecasting the price of the other. Greater risk for the suppliers may translate among other things into higher energy prices and reserve procurement costs. Furthermore, forecast errors may then cause inefficient offer and production decisions.

In the US model, however, the spot energy market and the operating reserve market are cleared simultaneously. The spot market-clearing algorithm minimises the cost of matching load under the constraint that enough spare generation capacity be available to provide reserve at all times. The market-clearing algorithm calculates the clearing price of energy as well as of each type of operating reserves. Each MW scheduled for production receives the energy clearing price and each MW supplying operating reserve is paid the clearing price of the operating reserve. The clearing price of operating reserves represents the marginal opportunity cost of maintaining the required amount of spare capacity.

NOTES

1. In fact, the value of electricity can vary widely from one minute to the next. For the purpose of trading, however, the standard electricity product is commonly defined in terms of total production, or consumption, during a fixed hour.
2. Examples of power exchanges operating on the spot and derivative markets are APX-ENDEX and N2EX in the UK, and APX-ENDEX in the Netherlands and Belgium; EPEX-SPOT and EEX operate, respectively, the spot and derivative markets in Germany and France.
3. See next subsection where we discuss complex offers.
4. If a vertical segment of demand and a vertical segment of supply overlap, the market-clearing price is undetermined. In that case, selection of the clearing price follows conventional rules that may vary in different markets.
5. In this section we draw heavily on Baldick, R., 2009. 'Single Clearing Price in Electricity Markets'; Report, February, (available at: http://www.competecoalition.com/files/Baldick%20study.pdf), which provides a comprehensive survey of the debate on the desirable auction format for wholesale electricity markets.
6. Pay-as-bid is not implemented in spot electricity markets. However, it is implemented in some European balancing markets; see Section 2.3.
7. Fabra, N., Von der Fehr, N. and Harbord, D., 2006. 'Designing Electricity Auctions', *RAND Journal of Economics*, **37**(1), 23–46; Federico, G. and Rahman, D., 2003. 'Bidding in an Electricity Pay-as-Bid Auction', *Journal of Regulatory Economics*, **24**(2), 175–211; Bunn, D. and Bower, J., 2001. 'Experimental Analysis of the Efficiency of Uniform Price versus Discriminatory Auctions in the England and Wales Electricity Market', *Journal of Economic Dynamics and Control*, **25**(3–4), 561–92; Rassenti, S., Smith, V. and Wilson, B., 2002. 'Discriminatory Price Auctions in Electricity Markets: Low Volatility at the Expense of High Price Levels', *Journal of Regulatory Economics*, **23**(2), 109–123; Cramton, P. and Stoft, S., 2007. 'Why We Need to Stick with Uniform-Price Auctions in Electricity Markets', *Electricity Journal*, **20**(1), 26–37.
8. We are abstracting here from the impact of the generator's inter-temporal constraints on the optimal bidding strategy. We discuss how these constraints are addressed by different market types later in this chapter.
9. These measures are discussed in Chapter 5.
10. The available generation capacity can be used either to produce electricity or to provide operating reserve. Therefore, generation scarcity must be assessed by comparing the available capacity with the sum of energy and operating reserve. For the sake of simplicity of exposition, in this section we refer only to the demand for energy. In Section 2.3.3 we address the relationship between the market-clearing prices for electricity and for operating reserve.
11. Note that we assess market demand and supply with reference only to physical resources. In other terms our (competitive) supply function is the system generators' variable cost function, and our demand function represents power withdrawals from the delivery points. In most European markets, bids and offers in the spot markets are less physical, as they refer to the net portfolio position of the market participants. This explains why in those spot power markets price-dependent demand bids are commonly observed. To the extent that there is perfect foresight and no impediments to trading, this feature of market design does not impact on the market outcome (quantity injected and withdrawn and spot prices), which will ultimately reflect the fundamental physical conditions of demand and supply.

12. Some large consumers have the capacity and may find it profitable to reduce consumption in response to price spikes. Typically those consumers find it more profitable to sell their availability to reduce consumption at short notice as an ancillary service or in the balancing market (see Section 2.3). In Figure 2.5 one may think of the consumption by those consumers in normal conditions as concurring to determine the price-inelastic demand function. The price-dependent load reduction by those consumers may instead be represented as generation capacity.

13. At this stage, disconnection is still a possibility. Whether disconnections will be implemented or not depends on the actual system conditions in real time.

14. Council of European Energy Regulators – CEER, 2005. *Third Benchmarking Report on Quality of Electricity Supply*.

15. Note that the problem has the same nature as the one caused by the pay-as-bid market-clearing rule.

16. The eight-hour block would have higher average costs since the start-up cost would be spread over a smaller quantity.

17. The power exchange operating the derivative market in France and Germany (EEX) also functions as clearing for OTC forward trading (the volume traded in 2010 was around 700 TWh).

18. Electricity futures are sometimes referred to as 'two-way contracts for differences', and options as 'one-way contracts for differences'.

19. As an alternative to generation or consumption, market participants may fulfil their delivery and consumption obligations by scheduling exports or imports with neighbouring countries.

20. In most markets, if a market participant fails to offset its contractual position with physical notifications, the system operator closes the market participant's position on its behalf.

21. The supply contracts to end-consumers are not regarded as transactions at a wholesale level. The client's consumption adds to the retailer's (negative) physical position. The retailer matches that position either by notifying own production, as the utility in our example, or by a purchase on the wholesale market, as the retailer in our example.

22. Notice that the trader's contract and physical positions are identical to those of a power exchange, since the same total quantity is bought and sold during each market session.

23. For example, the efficiency of thermal generator efficiency is affected by the ambient temperature.

24. If imbalances are assessed with reference to the sum of all production and consumption in the market participant's portfolio there is no need to require notification of a separate production or consumption programme per each generating unit owned by the market participants or for each consumer supplied by the market participant. However, this information becomes relevant if imbalances are assessed at a deeper level of granularity and if imbalances in opposite directions are settled at different prices, as we discuss later in this section.

25. The system operator buys and sells electricity on the real-time market continuously, based on the condition of the system at a given time. It is possible that during an hour the system imbalance may change from negative to positive and vice versa, causing the system operator to buy additional production and then to buy reduced production, for example.

26. See Section 2.3.4 and Chapter 4.

27. Since the British real-time market operates under a pay-as-bid pricing rule, the prices of multiple transactions that take place in the same 30-minute interval need to be aggregated into one System Buy Price and System Sell Price. The System Sell Price is computed as the average of the 500 MWh highest-priced offers accepted in the time interval required for assessing the imbalances (a fixed half-hour in the British system). The System Buy Price is obtained as the average of the 500 MWh lowest-priced bids accepted in the 30-minute time interval.

28. Vandezande, L., Meeus, L., Belmans, R., Saguan, M. and Glachant, J.-M., 2010. 'Well-functioning Balancing Markets: A Prerequisite for Wind Power Integration', *Energy Policy*, **38**(7), 3146–54.

29. More recent meters installed at residential consumers' homes record the volume of energy consumed in given time bands. For example, consumption occurred between 8.00 hours and 22.00 hours and between 22.00 hours and 8.00 hours. The roll-out of more advanced smart meters enabling real-time metering is being widely discussed.

30. For reasons of simplicity, we abstract, a number of issues that complicate the mechanisms actually implemented. These include, for example, the need to account for non-hourly metered generators, the fact that non-hourly meters are not all read simultaneously, the possibility of measuring the aggregate hourly consumption for each group of non-hourly metered consumers connected to each branch of the transmission network, and the implementation of a differentiated sharing factor, at different times of the day, in order to reflect the typical time profile of consumption by different categories of customers.

31. The benefit of the consumption shift would be split between all the suppliers of load-profiled customers, as the cost of the total load would decrease.

32. For reasons of simplicity, unless otherwise stated, our presentation assumes that only generators provide balancing services.

33. A call in the balancing market to increase or to reduce output is considered a change in the contractual position of the generator from its position at gate closure. As a consequence, a unit that does not achieve the production level after its balancing offer has been accepted incurs imbalance charges.

34. Arbitrage through voluntary imbalances can be limited by the structure of the imbalance prices in order to encourage market participants to balance their position at gate closure, as we discuss in Section 2.4.

35. Major electricity consumers may also supply balancing provided they are capable of reducing electricity withdrawals at the system operator's instruction.

36. Generators providing secondary reserves also need to be equipped with devices allowing production to vary automatically in response to a signal sent by the system operator.

37. In this case each offer on the capacity market also specifies the price of electricity if the capacity is activated. If the offer is accepted then that energy price will be offered on the balancing market. Primary reserve injections and injection reductions typically balance out over very short timeframes; such injections are typically not remunerated. The price for the electricity produced when the secondary reserve capacity is activated can either be fixed when the reserve offers are selected, or set as equal to the price prevailing on the balancing market in real time.

38. In the case of reserve capacity for downward regulation, a generator's energy schedule needs to be above minimum output, thus providing foot room.

39. The alternative approach is mentioned in the next section and analysed fully in Chapter 4.

40. Loads may also be re-dispatched, even though that is typically not the least-cost solution available to the system operator.

41. Real-world transmission networks often have a complex topology featuring multiple loops and parallel paths connecting any two points in the network. In such networks, binding network constraints may create relations among transactions across different locations with various degrees of complementarity or substitution. For example, if a transaction between two nodes is limited by a transmission constraint, this constraint can be relieved to a different degree by increasing net injections at some nodes of the network or by decreasing net injections at some other nodes. See Chapter 4.

42. The detailed approach and criteria for bid/offer exclusion from the balancing price calculation is described in the System Management Action Flagging Methodology (SMAF).

3. Generation capacity adequacy

**Guido Cervigni, Andrea Commisso and
Dmitri Perekhodtsev**

3.1 INTRODUCTION

In the market design discussed so far, known as an 'energy-only' market, generators obtain revenues (only) from selling electricity and ancillary services. In this context, persistently high electricity and ancillary service prices are relied upon to attract investment in generation capacity when the existing capacity is below the equilibrium level. Conversely, low electricity and ancillary service prices discourage capital accumulation when installed capacity is above the equilibrium level.

As in most other markets, the level of installed production capacity in energy-only markets is determined by the interaction between demand and supply of the final products supplied. This differs substantially from the traditional approach, in which utilities meet reliability and resource adequacy requirements according to engineering standards regarding the acceptable hours of load shedding, based on the expected load variance and generator availability. In the energy-only design, market forces rather than engineering standards determine the installed capacity, and ultimately the level of reliability.

A pure energy-only market design is difficult to implement for both political and technical reasons. First, an energy-only market is characterised by generally moderate energy prices with rare price spikes. Occasional capacity shortages and extremely high scarcity-related prices are a normal feature of a well-functioning energy-only market. Policy makers and regulators are generally unwilling to accept the potentially severe price spikes and the instances of demand rationing (which may include rolling blackouts) associated with energy-only markets.

Second, wholesale electricity markets are particularly vulnerable to market power when existing capacity is close to full utilisation. Market-power mitigation mechanisms may intentionally or unintentionally reduce the price for electricity and ancillary services during conditions of scarcity. If this happens, the optimal level of generation capacity cannot operate profitably.

Finally, in energy-only markets a large portion of some generators' fixed costs is covered by the revenues obtained during genuine conditions of scarcity. This makes the investment in generation capacity risky, since even small changes in the number of scarcity events can have a dramatic impact on the producers' revenues. The problem is exacerbated by the fact that in the event of scarcity the price needs to be set administratively, because electricity demand is largely insensitive to price. To some extent even the detection of situations of scarcity may be impaired by some features of the market design.

These issues, combined with the long construction times of generation plants, result in a boom-and-bust pattern in generation investment, and governments and regulators are concerned that during the low phases of investment cycles installed generation capacity may not be sufficient to match the load at all times.

Almost all power markets implement forms of out-of-market backstop mechanisms in order to ensure reliability and sufficient generation capacity to match load. In addition, explicit capacity support schemes have been implemented in the US markets, and in Spain and Italy in Europe. Their introduction is currently under discussion in the UK, France and Germany.

In Section 3.2 we assess how the specific features of electricity impact on the economic mechanism driving investments in generation capacity, and investigate why this may create the need for capacity support schemes. In Section 3.3 we analyse alternative capacity support schemes.

Throughout the chapter we maintain the assumption of perfect competition in the electricity and ancillary service markets; we specifically assume that entry and exit on the market are frictionless. This ensures that – at equilibrium – investment in generation capacity yields no more than the minimum rate of return necessary to attract capital to the industry. Under this assumption, capacity support schemes do not increase the return of investment in power generation, but only the level of installed capacity.

3.2   THE RATIONALE FOR GENERATION CAPACITY SUPPORT SCHEMES

In this section we assess how the specific features of electricity impact on the economic mechanism driving investment in generation capacity, and investigate why this may create the need for capacity support schemes.

Capacity adequacy concerns mostly relate to distortions in the market outcome under conditions of scarcity. Scarcity hours are particularly important in the electricity industry because a large portion of some generators' fixed costs must be recovered during these hours. In fact the generating unit with the highest variable cost installed in the system, the marginal unit, should cover its entire fixed cost by producing during scarcity hours, as it is only during these hours – under perfect competition – that the market price is greater than the variable cost of the marginal generator. For this reason, even moderate distortions of the electricity prices prevailing during scarcity hours, or in the number of scarcity hours, could have a major impact on the generators' profitability.

We identified three broad motivations supporting the introduction of capacity support schemes. First, a large part of electricity demand is currently price inflexible in the short run. When the price-insensitive portion of demand exceeds available generation capacity, involuntary load reduction via disconnections, or load shedding, may become necessary, as we discussed extensively in Section 2.2.1. In this case the price for electricity must be administratively set. Imperfections in the administrative process that sets the market price for electricity in the event of scarcity may bias the incentives to invest in generation capacity.

Second, some features of the market design and regulatory system may prevent energy and operating reserve prices from rising to levels that correctly reflect conditions of scarcity. In this case the generation capacity is under-remunerated in scarcity situations, which results in underinvestment.

Third, capacity adequacy concerns are sometimes motivated by the specific risk structure of the generation business, such that small changes in demand or supply conditions can have a dramatic impact on generators' profitability at times of scarcity. While the first two issues call for mechanisms that integrate the generators' income in order to attract an efficient level of investment, the third issue can be handled by coordinating the timing of investments in generation capacity in order to reduce the risk for investors. A more certain environment is expected to reduce the rate of return required by investors, to the ultimate benefit of consumers.

We discuss possible flaws of the price-setting mechanism in the event of scarcity in Section 3.2.1, the 'missing money' problem in Section 3.2.2 and the coordination role of capacity support schemes in Section 3.2.3.

3.2.1 Flaws in the Assessment of the Value of Lost Load

Figure 3.1 shows the wholesale electricity market equilibrium for two sets of hourly demands with different price elasticities. The price-insensitive demand results in fewer scarcity hours, that is, hours when the market–clearing price rises above the system's marginal cost (SMC), the marginal cost of the most expensive generating unit in order to ration demand.

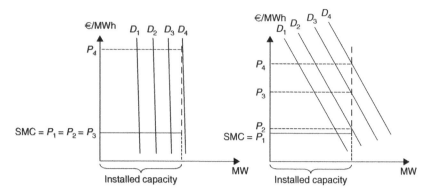

Figure 3.1   Market equilibrium with different demand elasticity

However, during those hours the distance of the clearing price from the system's marginal cost is greater than it would be if demand were more flexible.

In the current situation a large portion of the electricity demand is completely price insensitive in the day-ahead to real-time timeframe. This reflects the technical features of most of the metering systems currently in place. The meters installed at small consumers' premises typically record the consumers' total withdrawal over long periods, typically one month. Older meters record only the total consumption since the equipment was installed. It is therefore impossible to measure the volume of electricity withdrawn by a consumer during each hour. When hourly consumption is not known, retail prices cannot reflect wholesale market prices, and therefore cannot signal to consumers the cost of their consumption in each hour.

Without demand response, no matter how high (wholesale) prices rise, load will not reduce to the level of available generation capacity during scarcity events. Therefore a regulatory solution for scarcity pricing and involuntary load reductions must be implemented in order to avoid uncontrolled widespread service disruptions. Given that selective disconnection is technically infeasible, if necessary all the consumers connected to the same network branch will be disconnected at the same time.[1]

As such quantity-rationing events have no associated market-clearing price, the price for electricity in scarcity hours must be set administratively. The appropriate regulated price in such circumstances is the estimated value of lost load, or VoLL. The VoLL is based on an estimate of the amount that customers would be willing to pay in order to avoid being disconnected. In other words, the VoLL is supposed to be the price that makes consumers indifferent between consuming electricity at that price and not consuming. The VoLL is typically found to be several orders of magnitude greater than average electricity prices. VoLLs in the range of €5,000–€10,000 are commonly regarded as plausible.

Implementing VoLL pricing and load curtailment is not without problems. First, load curtailment is perceived by end–consumers as being unfair. Curtailed consumers typically do not receive payments equivalent to the VoLL from their suppliers, while non-curtailed consumers are not charged for the VoLL.

Second, although each consumer might give a different value to electricity, current technology makes it impossible to selectively disconnect consumers based on their individual valuation of electricity. It is then impossible to provide incentives to consumers to reveal their individual valuation.

Finally, load shedding and price spikes rapidly become a matter for political concern.

For these reasons governments and regulators and system operators tend to pursue more or less explicit capacity targets, rather than relying on extremely high prices under conditions of scarcity in order to attract investment in generation.

3.2.2 The Missing Money Problem

The missing money problem occurs when some elements of the market design, industry regulation or industry practices cause generators' revenues to be systematically insufficient to attract the efficient level of investment. When revenue deficiency becomes a structural feature of the market, the result is a drop in installed capacity.

Capacity support mechanisms are therefore intended to integrate generators' income. Below we discuss the potential causes of missing money.

Market-power mitigation measures
As we discuss in detail in Chapter 5, wholesale electricity markets are particularly vulnerable to the exercise of market power when existing capacity is close to full utilisation. When demand approaches the level of available capacity, even relatively small generators enjoy market power. Since both electricity supply and demand are to a large extent price inflexible, withdrawing even a small amount of capacity from the market when the system is tight can be very profitable for a generator, as it may result in a dramatic increase of the market-clearing price. This happens especially if capacity withdrawal results in a scarcity situation, that is, if the market-clearing price jumps from the system marginal cost to the much higher VoLL.

A straightforward solution to reduce generators' incentive to withdraw

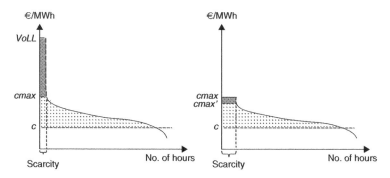

Figure 3.2   Price duration curves

capacity when the system is tight consists of setting the scarcity price at a level much below the VoLL. For example, the scarcity price could be administratively set as equivalent to the marginal cost of the most expensive generating unit activated. With such a price cap, the expected revenue for the generators is below what is necessary to attract an efficient level of investment. This situation is illustrated in Figure 3.2, where we show a price duration curve. The price duration curve shows, for each price level, the number of hours the market-clearing price is above that level.

In the figure, the area with the grey background represents the profit that 1 MW of generation capacity with variable cost c obtains on the wholesale market. The profit is equivalent to the sum of the difference between the market price and the generator's variable cost in each hour. Over the lifetime of the generator, capacity settles to the efficient level and composition, and market prices are such that the efficient capacity level obtains the standard return on investment.

If, during conditions of scarcity, the price is set to be equivalent to the most expensive generator's cost (*cmax* in the figure) instead of the VoLL, the generators' profits are reduced by the darker area.[2] Note that because of the price cap, the most expensive unit does not receive any contribution to fixed costs from selling energy. In the long run, standard profitability conditions are re-established via the entry/exit process. Capacity settles at a lower level as the units with variable cost *cmax* are not replaced, and the new system marginal cost becomes *cmax'*. This means that the number of scarcity hours increases until the new dark area is large enough to cover the generators' fixed costs,[3] as shown in the right panel of Figure 3.2.

When the scarcity price is below the VoLL, a mechanism integrating the generators' revenues is necessary to ensure that the efficient capacity level is available in the system at all times.

In most markets, price caps below common estimates of VoLL are imposed. This happens, for example, in Nordpool (the Nordic power market), in the Australian day-ahead market and in the US in ERCOT (the power market operating in Texas). The Australian market and ERCOT also have additional price mitigation measures which limit the duration of elevated scarcity prices. If prices remain above a predefined threshold for a certain period of time a price cap is enforced in Australia, and the normal price cap is lowered in ERCOT.

An alternative approach to market power mitigation, discussed in Chapter 5, Section 5.3, involves capping the generators' offers only at times when they are considered to enjoy significant market power. However, when the system is tight it is very hard to distinguish between high prices that reflect a genuine situation of scarcity and high prices that are the result of exercise of market power. This was particularly evident in the aftermath of the 2000–01 California power crises where, according to some observers, high loads and low water availability for electricity production, combined with market manipulation, resulted in extremely high prices. Market-power mitigation mechanisms based on selective capping of the generators' offers may also be activated in situations of genuine scarcity and create revenue deficiencies for the generators.

For this reason, such measures are typically associated with capacity support mechanisms providing an additional source of revenues to the generators. The class of capacity support schemes presented in Section 3.3.1 accomplishes both functions: limiting generators' revenues during scarcity events, and making up the missing money with a payment for capacity availability over a longer period.

Out-of-market procurement of reserve services
Almost all power markets have developed out-of-market backstop mechanisms for ensuring reliability and sufficient capacity. In most cases, market operators simply procure reserve capacity outside the market framework if they expect peak capacity to be short of their targeted reliability standard. In Nordpool, for example, when capacity is forecast to be insufficient on a three-year forward basis to meet the reliability target, the system operator is authorised to procure peaking resources under long-term contracts with the costs of the procurement paid for by the state.[4] In the UK, operating reserve is procured by the system operator under long-term contracts of up to five years in order to provide sufficient investment signals to providers and allow enough time for the repayment of a provider's investment.[5] In some US markets, out-of-market capacity purchases have been made in the form of reliability-must-run contracts, which are intended to retain in the system capacity resources that might otherwise be retired or mothballed.

Such backstop measures may displace market-driven investment in generation capacity and inhibit the development of demand-response measures in energy-only markets if they prevent market prices rising to VoLL during scarcity events. As we discuss in Section 3.3.3, the distortions caused by out-of-market mechanisms can be overcome by appropriately designed scarcity pricing rules, administratively setting the market price to the VoLL each time the backstop resources are activated.

Lack of transparency

A number of markets do not implement scarcity pricing rules that administratively set the electricity price to VoLL when scarcity conditions are detected by the system operator. These markets rely on generators to increase their bid prices above marginal costs in order to set scarcity prices. For this mechanism to be effective, it is crucial that generators are in a position to correctly anticipate scarcity situations. Lack of information about the demand and supply may cause some scarcity situations to go undetected by the market participants, exacerbating the missing money problem.

When multiple related markets are cleared independently, the profit-maximising bid for a generator in one market depends on the expected equilibrium price in the others. Arbitrage between the related markets should result in all the markets clearing at a price consistent with the overall demand and supply. As discussed in Chapter 2, some elements of the designs implemented in Europe may make arbitrage between the various markets difficult: energy and reserve markets are cleared independently; the design of the day-ahead, intraday and real-time markets is not always homogeneous; and speculative trading against real-time prices is limited in some markets. Imperfect arbitrage may result in scarcity conditions not being reflected in the same way in different markets. Consider, for example, the case in which the energy market clears first and the operating reserve capacity market clears later. If the generators fail to anticipate the scarcity situation when formulating their bids in the energy market, the electricity clearing price will not signal scarcity because the available generation capacity is greater than the demand in the energy market. However, scarcity conditions will emerge – and result in high prices – in the operating reserve market. If this is the case, the generation capacity committed on the electricity market will receive a price that does not correctly reflect its value.

3.2.3   Coordination of Investment Decisions

The high level of risk in generation investments is sometimes mentioned as a reason for the introduction of capacity support mechanisms. In this respect, the relevant feature of some capacity support schemes is that they coordinate market participants' investment decisions. To the extent that such coordination reduces the uncertainty faced by generators, it also reduces the required rate of return on the investment in generation capacity. As a consequence, all other things being equal, a higher level of capacity will be installed.

In order to illustrate this approach, consider an investor assessing the opportunity to invest in a 1,000 MW plant to be in service in year t. The profitability of this investment crucially depends on the decisions of other potential investors to enter the market. If just two new plants were brought into service in year t instead of only one, electricity prices might turn out much lower than if only one were built. The impact of the second project on market prices could be considerable for several years, potentially undermining the profitability of both projects.

Although an efficient investment pattern would include 1,000 MW additional capacity in service from year t, each investor will want to reduce its investment's vulnerability to other investors' decisions. Such a strategy can be expected to lead to investments being delayed when compared with the efficient path. If this happened, an inefficiently low level of available capacity would be available in year t.

In this context, a mechanism that coordinates investment decisions could be beneficial. A central entity could set a capacity target for each year, and select the parties that would make the target capacity available in exchange for payment. Selection of the new capacity provider could take place by means of auctions.

In our example, the target level at time t would be such that only the additional 1,000 MW would receive the capacity payment. If the central entity commits to paying for capacity availability well in advance, the scheme coordinates the decisions of the potential investors. Only one of the potential investors obtains the capacity payment for 1,000 MW capacity at time t. The other investors would be unlikely to sink money into making additional capacity available at time t, as they know that such an investment would lead to excess capacity overall, and would therefore be unprofitable.

Here the capacity support scheme is welfare improving because it coordinates the timing of the investors' decisions. While in the previous section the capacity support scheme increases the generators' income in order to compensate for the effects of the market flaws, here the capacity support scheme acts mainly as a coordination device. In our example, very little or no compensation would be required by the winner of the auction, if the central entity auctions off only 1,000 MW of incremental capacity. In this case, each of the potential investors would find it profitable to make

1,000 MW capacity at time *t* with no further compensation, provided that no more than 1,000 MW new capacity is built in total. What the auction process delivers is mainly the certainty as to who will make the investment. If instead, as typically happens, the capacity target pursued by the central entity is greater than the level that would attract the investment, then the auction will clear at a higher price.

Note also that the reallocation of risk from the generators to the central entity acting on behalf of the consumers is a byproduct of this measure, not the source of its expected welfare-improving effect.

3.3   CAPACITY SUPPORT SCHEMES

Three broad approaches to the design of capacity support schemes can be identified. The first approach sets the price that a central entity, on behalf of the consumers, commits to pay for all available capacity. Capacity payments add to the revenues obtained by generators from selling electricity and ancillary services; the higher expected income is supposed to attract additional investment in generation capacity. The second approach sets the volume of available capacity that the central entity commits to paying for, either directly or by placing an obligation on the load-serving entities. This creates the demand for a product, the available capacity, which generators can supply. The interaction between the regulatory-driven demand and the supply of available capacity determines the market-clearing price for the available capacity. The third approach consists of reserving a certain generation capacity for use only in scarcity situations, as a substitute for load curtailment.

In the rest of this section we discuss each approach in turn.

3.3.1   Capacity Payments

Capacity payments are administratively set payments per MW for available capacity, regardless of whether it is dispatched to run. Capacity payments are intended to provide generators with additional revenues equivalent to the missing money. Consider, for example, a market where the missing money issue is created by an overall price cap equivalent to the variable cost of the marginal generator, typically an open-cycle gas-fired unit. As illustrated in Section 3.2.2, the efficient level of capacity would be under-remunerated because of the cap. Each MW of installed capacity would, over its lifetime, miss out on revenues equivalent to the fixed cost of the marginal unit, the variable cost of which sets the price cap. In this case the efficient capacity payment would equal the fixed cost of the open-cycle gas-fired unit. This would neutralise the impact of the price cap on the generators' income,[6] and therefore re-establish the correct investment incentives. The level and allocation of capacity payments is a matter of administrative judgement.

Different schemes feature different availability requirements. For illustration purposes we shall consider two extreme methods. The first methodology would pay 1/8,760 of the annual fixed cost of the marginal unit for all capacity that turns out to be available during each hour of the year.[7] The drawback of this approach is that it does not provide incentives to make capacity available when the system needs it most; the generator has the same incentive to be in service all hours, independently of the level of demand.

The second extreme methodology would pay $1/N$ of the annual fixed cost of the marginal unit for all the capacity that turns out to be available in each the N hours of the year when the system operator expects scarcity. The advantage of this approach is that it provides stronger incentives for generators to make capacity available when the system needs it most. If the N hours selected by the system operator, and only those N hours, turn out to be scarcity hours, this methodology provides exactly the same incentives that would be provided by pricing electricity at the demand-rationing VoLL price. The scheme, however, has an additional advantage over VoLL scarcity pricing: it removes the burden of predicting when scarcity will occur from generators. For example, they can plan maintenance outages at times other than those when the capacity payment is granted. The drawback of this approach is that it depends on the system operator's ability to predict exactly when scarcity conditions will occur. Typically, critical system conditions could manifest 5–20 hours per year, and predicting, say, a year in advance when those hours will be is a difficult exercise.

The trade-off between the power of the incentives to make capacity available and the risk of excluding some scarcity hours from the scope of the mechanism is addressed in practice by granting capacity payments in exchange for availability for a relatively large subset of a year's hours, for example one or two thousand, when demand is expected to be high. In Chile, for example, capacity payments are granted for availability in the months May–September; in Colombia, in the dry December–April season when hydropower production is limited.[8] In Italy the set of critical days when capacity availability will be remunerated is set yearly by the system operator.

Administratively determined capacity payments can target new resources. The Spanish capacity payment system, for example, comprises a component granted only to new capacity. This approach is sometimes supported in policy discussions, because it reduces the initial amount of capacity payments compared with capacity payments being granted for the entire capacity. However, it is distortive, since the new capacity attracted by the selective capacity payment will exacerbate the missing money problem for existing generators that do not receive the capacity payment. This will accelerate substitution of the generating fleet.

The British pool system, operational between 1991 and 2000, featured a capacity payment component that was paid to all generators available for dispatch, irrespective of their activation. The capacity payment was computed the day before delivery as the expected value, over the probability distribution function of the demand realisations, of the scarcity rent on the day of delivery. As a result, capacity payments would be low when available capacity was high compared with load, and payments increased as the reserve margins shrank.[9] The design of the early UK capacity payment mechanism can be interpreted, more than as a capacity support system, as a way to implement a pure energy-only market under the constraint that the real-time price for electricity and operating reserve be fixed one day in advance. However, the parameters of the mechanisms were set in a way such that it resulted in generous payments to the generators.[10]

3.3.2 Capacity Requirements

With capacity payments, available capacity levels remain uncertain, since they depend on the market response to administratively set prices. An alternative approach is to ensure resource adequacy by imposing a reserve margin requirement on all electricity retailers. In this section we discuss two support mechanisms based on capacity requirements that differ in the content of the obligation placed on capacity suppliers.

Reserve requirements
In this approach the system operator sets the capacity requirement. The required level of installed capacity is generally set around 115–118 per cent of the peak load, a figure derived from engineering standards. Like the level of the capacity payments, reserve requirements are set administratively, and the trade-off between the cost of achieving the reliability target and the value provided by that reliability is typically not explicitly addressed.

The reserve requirement is then divided between retail suppliers in proportion to the expected contribution of their clients to peak load. Each retail supplier is responsible for acquiring capacity rights that exceed its predicted peak load by the required reserve margin, either through self-supply or by contracting available capacity from generators. The capacity availability contracts can be negotiated either bilaterally or on organised markets.

By selling capacity, a generator commits to make the contractual volume of capacity available during the period of the contract. The obligation is fulfilled independently of the actual use of that capacity. The generator can use that capacity to deliver electricity sold bilaterally, on the spot market or on the real-time market. Even if the capacity turns out to not be used, the obligation has been fulfilled so long as it has been offered as operating reserve and on the real-time market. Financial penalties apply if the capacity is not delivered. Resources wishing to supply capacity apply to the system operator to be assigned capacity credits, typically based on their historic availability record; underdelivery can also be penalized through a reduction in the capacity credits assigned in the future.

Placing a capacity requirement on retailers creates the demand for capacity that meets the generators' supply. A market for capacity is then established. The price in that market settles at the level that attracts investment in generation capacity up to the system operator's requirement. Under the usual perfect competition assumptions, generators' revenues from selling capacity availability are equivalent to the missing money.

Mechanisms based on capacity requirements are extensively implemented in US markets. Earlier capacity requirement systems were implemented in the context of traditionally regulated markets, where integrated utilities carried a regulatory obligation to procure the generating capacity needed to meet the resource requirements in their (exclusive) service areas. The absence of retail competition allowed the utilities to recover the costs associated with that obligation through regulated retail rates. That meant that the need for adjusting the utilities' capacity portfolios via trading was limited to transitory imbalances, since the resource planning requirements were overseen or enforced by the state regulators.

In restructured markets, with retail competition and small retailers, some features of the older mechanisms have caused concern. In particular, enforcing the capacity requirement just days or months before the relevant delivery period may lead to extreme price volatility, with capacity prices jumping from the cap (when there is insufficient capacity) to zero (when there is excess capacity). This happens because in such a short time horizon both the demand for and the supply of capacity are highly inelastic. The supply of capacity to deliver within a period of months does not include units not yet built. As a consequence, the entrant's cost does not act as a ceiling to the prices of capacity.

In addition, suppliers and possibly buyers of capacity contracts may enjoy significant market power when the system is close to the target resource requirement: suppliers may be able to move price from close to zero to the cap even by withholding relatively small amounts of capacity. Buyers may similarly be able to move capacity price levels close to zero by slightly reducing their demand for capacity, for example by declaring that they will meet part of their obligation via self-supply. Finally, if capacity deficiencies are detected only slightly in advance, it may be impossible or extremely costly to the system operator to make up the missing resources.

In order to address those concerns, forward reserve requirements have been introduced in several US power markets. For example in PJM (the power market in Pennsylvania, New Jersey and Maryland) and in ISO-NE (the power market in New England) load-serving entities are required to procure sufficient resources for up to three years ahead of the delivery year. In addition, both PJM and ISO-NE allow some suppliers of new capacity to lock in capacity prices for three to five years.

Requiring resource commitments sufficiently in advance of delivery leaves enough time for either market participants or the system operator to procure additional resources if deficiencies are detected. The time horizon also gives capacity suppliers enough time to modify their resource development plans, for example by bringing mothballed plants back online, making the capital investments necessary to defer the retirement of other plants, speeding up the development of a new power plant, or developing additional demand response capabilities. As a result, the price elasticity of the capacity supply curve rises, price volatility is reduced and competition increases.

In the PJM and NYISO (the power market in the state of New York) markets a certain degree of price elasticity is also introduced on the demand side by implementing a downward-sloping capacity demand curve, which varies the resource adequacy requirements as a function of capacity prices. As we have already shown in Figure 3.1, a flatter capacity demand curve reduces price volatility, as a shift in the demand or supply curve leads to a smaller change in the market-clearing price. However, the capacity demand curve implemented in PJM and NYISO is not intended to represent the consumers' (estimated) willingness to pay for capacity. Instead, consumers are assumed to be available to pay a price equivalent to the estimated building cost of new peaking resources for an administratively set target level of capacity. Then the slope of the curve near the target level of capacity is based on an administrative judgement As a result, rather than reflecting consumer preferences, the capacity demand curve basically implements a cost-based cap on the price of capacity.

In order to address transmission congestion issues, the capacity requirements are imposed on a zonal or locational basis in some markets, including PJM and NYISO.

With retail liberalisation, the number of retailers trading capacity contracts has increased. In addition, the possibility for consumers to switch supplier creates an additional need for trading capacity contracts, in order to adjust each retailer's capacity portfolio to the changing customer base. In order to address this development, centralised capacity exchanges run by the system operators have been introduced in several US markets. Centralised markets provide a backstop procurement mechanism, reduce transactions costs and provide greater liquidity and pricing transparency.

Energy options backed by generation capacity
Administratively set capacity targets may also be achieved through financial contracts with the suppliers of capacity. Capacity support schemes based on financial obligations are implemented in Colombia – the 'firm energy obligation'. In Europe a mechanism along the same lines is being introduced in Italy.

In this methodology, the capacity contract takes the form of a call option on the generators' capacity. In exchange for a fixed fee, the supplier of generation capacity commits to pay the counterparty the following amount in each hour of the contract period:

$$\text{Max}(0, p_t^{Spot} - p^{Strike}),$$

where p_t^{Spot} is the spot price of electricity in the hour, and p^{Strike} is the option's strike price, set as equal to the variable cost of the marginal generation unit in the system. In scarcity hours, when the price rises above the variable cost of the marginal unit, the contracted generator disburses the scarcity rent for each MW of hedged capacity. The scarcity rent is equivalent to the difference between the market price and the system marginal cost, that is, the variable cost of the marginal generator.

This provides the correct incentive for generators to make the hedged level of capacity available when the capacity is most valuable to the system. Consider the situation of a generator that has sold a 1 MW capacity contract and has not made the corresponding capacity available in a scarcity hour, when the electricity price equals VoLL. The generator suffers a loss on the capacity contract equal to $p_t^{Spot} - p^{Strike}$, which amounts to the scarcity rent in scarcity conditions. The generator can offset this loss by making available 1 MW capacity in that hour; by doing so the generator sells 1 MWh and obtains profit $p_t^{Spot} - VC$, where VC is the generator's variable cost. The net profit to the generator is then equivalent to the payment due under the capacity contract:

$$-(p_t^{Spot} - p^{Strike}) + (p_t^{Spot} - VC) = (p^{Strike} - VC).$$

A further advantage of this methodology is that it mitigates market power by capping the net revenues for contracted capacity at the system marginal cost. However, while it caps generator revenues, the mechanism does not cap the market-clearing price, which will rise to the level necessary to ration demand when the system is tight. As a result, the incentives to develop demand response resources are not distorted as they would be in the case of a price cap.[11]

In the Colombian implementation, additional provisions ensure that the financial obligation placed on the supplier of capacity is backed by physical generation capacity, and that the capacity is operated in such a way as to ensure its availability at scarcity times. For example, contracted thermal generators must provide proof of fuel availability during the commitment period.

Like installed capacity requirements, energy options backed by generation capacity may be procured directly by the system operator, or by placing an obligation on the load-serving entities. The options may have different duration, and may be procured more or less in advance of the commitment period. Finally, the option's strike prices may be fixed or indexed.

In Colombia, the system operator sets the capacity requirement and procures the corresponding energy options three years ahead of the start of the commitment period, which ranges from one to 20 years. The energy options are procured via auctions where generators are selected according to their bids for the option's fixed fee. The option's strike prices are indexed to the price of natural gas, the fuel firing the marginal open-cycle gas turbine (OCGT) capacity.

Capacity support mechanisms based on capacity requirements may contribute to coordinating generation capacity investment decisions, provided that the capacity contracts are awarded well in advance of the time of delivery. In this case a would-be investor may make the decision to build new capacity conditional on the auction's outcome. The auction outcome conveys important information to the participants. Broadly speaking, the auction's winners:

- learn that they are probably more efficient (or less risk-averse) than the others; and
- hedge part of their revenues.

Would-be investors not awarded capacity contracts in the auction learn that:

- they are probably less efficient than the winners;
- the system operator's capacity requirement will be covered by the winners of the auction; and

● in the event that they decided to go ahead with the investment, they would have no protection against excess-capacity situations, leading to too few scarcity hours.

The information conveyed by the capacity support scheme coordinates the investors' decisions. The decision to invest or not to invest is made easier. The auction's winners face strong incentives to invest, since they can be confident that the auction's losers will not invest and bring about excess capacity; the auction losers have strong incentives not to invest, since they can be confident that they will not miss a profit opportunity.

3.3.3 Reserve of Last Resort

This approach, implemented in Sweden and Finland,[12] is based on reserving part of the installed generation capacity for use only in scarcity situations, that is, as the reserve of last resort.

In order for the measure to bring about a permanent increase of installed generation capacity, the last-resort reserve has to be effectively removed from the market. This means that each time the last-resort reserve is activated, the market price for electricity (and operating reserve) rises to the VoLL, as in the event of scarcity. Otherwise, as we discussed in Section 3.2.2, the reserve of last resort will displace new capacity and the total installed capacity will not increase.

Consider, for example, the market shown in Figure 3.3, where we assume that demand and installed capacity are steady. Assume that the regulator is not satisfied that the current level of installed capacity is adequate and believes that an additional capacity of 2,000 MW is necessary.

In order to induce investment in an additional capacity of 2,000 MW, the regulator therefore contracts 2,000 MW of the existing capacity as last-resort reserve. The contracted capacity is then offered on the energy

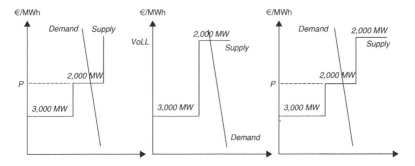

Figure 3.3   Reserve-of-last-resort scheme

and ancillary service markets at a price equal to the VoLL. The result of this measure is the market supply function represented at the centre of Figure 3.3. Consequently, in the event that the last-resort capacity is scheduled for production or to provide operating reserve the market-clearing price is the same as in the event of scarcity.

The new aggregate supply function is such that the market-clearing prices reflect scarcity conditions more often than they would without the intervention. The profitability of the existing generation capacity then increases. This attracts investment until the total installed capacity reaches the pre-intervention level, that is, until the capacity shifted to the last-resort reserve has been replaced. The new equilibrium is shown in the right panel of Figure 3.3.

The scheme based on the reserve of last resort may cause inefficiency if the units providing reserve of last resort turn out to be more efficient than some other units. In this case, cheaper units are withdrawn from the market while more expensive generators are activated to meet load. For this reason, the reserve of last resort appears particularly attractive when the regulator has the opportunity for preventing old and inefficient units being scrapped. The cost of keeping alive units that would otherwise be dismantled could be relatively low, and there would be little risk of technical inefficiency.

NOTES

1. The deployment of smart meters could in the future overcome some of the issues discussed in this section by allowing recording of hourly consumption, and therefore implementation of hourly differentiated retail prices. Alternatively, smart meters fitted with remotely controllable switches could make it possible to selectively disconnect consumers that state less willingness to pay, while charging the others the rationing price.
2. In the example we ignore the revenues that generators obtain by providing ancillary services, and we refer to electricity spot-market sessions only. These simplifications are irrelevant provided that the same price cap is consistently enforced on all services and market sessions.
3. We have assumed that the price cap has not been adjusted to the new (and lower) system marginal cost. If this happened, the installed capacity would continue to shrink.
4. See Nordel, 2007. *Guidelines for Implementation of Transitional Peak Load Arrangements: Proposal of Nordel*, available at: http://www.svk.se/Global/01_Om_oss/Pdf/Elmarknadsradet/071115NordelGuidelines.pdf.
5. See National Grid Electricity Transmission, 2008. *Long-Term Reliability Assessment, 2008–2017*, available at: http://www.nerc.com/page.php?cid=4|61.
6. The market-power mitigation effects of the cap still hold, as the capacity payment is independent of the generators' offers.
7. We ignore maintenance stops for reasons of simplicity. The correct assessment would allow each generator to obtain the annual fixed cost of the marginal unit in a number of hours equal to the difference between 8,760 – the number of hours in a year – and the duration of a standard maintenance period.

8. In Chile a penalty for failure to deliver capacity based on the VoLL re-establishes the correct incentives for the generator to make capacity available in the scarcity hours.
9. The capacity payment was computed as the product of the loss of load probability assessed by the system operator for the following day and the difference between the VoLL and the system marginal cost.
10. For a detailed analysis of this mechanism see, for example, Roques, F.A., Newbery, D.M., and Nuttall, W.J., 2005. 'Investment Incentives and Electricity Market Design: The British Experience', *Review of Network Economics*, **4**(2), 93–128.
11. A similar mechanism operates in ISO-NE, in the context of a capacity requirement system. The scarcity rent collected by the generator when the scarcity pricing mechanism is triggered is subtracted from the monthly capacity payments.
12. The RMR contracts implemented in some US markets, discussed in Section 3.2.2, are based on the same logic.

4. Congestion management and transmission rights

Dmitri Perekhodtsev and Guido Cervigni

4.1 INTRODUCTION

In this chapter we discuss the impact of electricity transmission technology on the design and outcome of electricity markets.

Electricity is transported on a transmission network from the place where it is generated to the place where it is used. Transmission congestion occurs when generation and demand schedules cleared in the market lead to a set of power flows violating one or more network constraints. When this happens, congestion must be alleviated. Electricity transmission is different from other goods and commodities for which transport flows can be re-routed relatively easily. On the contrary, increasing (net) injections at some locations and decreasing them at others is the primary and often only option to relieve congestion.

Market arrangements differ across electricity markets in the way they induce market participants to deviate from the injection/withdrawal levels they would implement in the absence of any transmission constraints. Two general approaches can be identified. The first restricts the transactions that market participants can enter into to those producing feasible power flows. This is achieved by allocating and enforcing a feasible set of rights to inject and withdraw power at the different network locations. As a result, in the event of congestion, demand and supply clear at different prices at different locations, reflecting the different costs of meeting the incremental demand for electricity at each location. Generally the market-clearing prices at import-constrained locations are higher than at export-constrained locations. This general approach has been adopted in many US electricity markets, such as PJM, New York, New England, California, Texas and the Midwest, as well as in New Zealand. In Europe this approach has been implemented in the markets of Norway, Italy, and recently Sweden, where several network nodes are grouped into market zones, and the set of feasible market transactions is the result of constraints on the net power flows between the zones. The ongoing projects aiming to integrate the European national spot markets also feature locational pricing, in which each participating country is treated as a market zone.

The second approach compensates market participants for deviating from the level of injections and withdrawals they would have selected in the absence of transmission constraints. This practice is termed 're-dispatch'. In this approach the electricity market is initially run neglecting any transmission constraints. Subsequently, if the market outcome produces power flows that violate one or more transmission constraints, generators and possibly consumers are paid to modify the level of injections and withdrawals they have scheduled in the first unconstrained stage. This approach is widely adopted in the UK, France, the Netherlands, Spain and Germany.

In Section 4.2 we describe how net injections at the nodes of a transmission system produce flows over elements of the transmission network. In Section 4.3 we discuss congestion management methodologies based on restricting the transactions that market participants can enter into to those producing feasible power flows. In Section 4.4 we discuss the alternative approach, in which market transactions are not subject to any network-related restrictions, and congestion is relieved at a subsequent stage. In Section 4.5 we discuss an intermediate approach, in which network-related constraints to market transactions are defined at a zonal level. In Section 4.6 we address the longstanding policy debate on the relative merits of congestion management via re-dispatch and via locational price differentiation. In Section 4.7 we discuss alternative approaches to transmission network development.

4.2 NETWORK EFFECTS AND LOCATIONAL DIFFERENTIATION OF THE VALUE OF ELECTRICITY

When one or more network constraints are binding, injections at different nodes may not be substituted or can only be partially substituted to meet a given demand. Real-world transmission networks often have a complex topology featuring multiple loops and parallel paths connecting any two points in the network. In such networks, binding network constraints may create relations between transactions across different locations with varying degrees of complementarity or substitution. For example, if a transaction between two nodes is restricted by a transmission constraint, this constraint can be variously relieved by increasing net injections at some nodes of the network or by decreasing net injections at some other nodes.

When network constraints are binding, the cost of meeting an incremental increase in demand can differ between nodes, as the least-cost way to meet the increased demand at different nodes may involve increasing the production of different generating units. In this section we analyse the effect of network constraints on the cost of meeting a net demand increase at the different network nodes, while in Sections 4.3, 4.4 and 4.5 we investigate how this feature of the electricity industry impacts on the market outcome, under alternative market designs.

In Section 4.2.1 we illustrate the relationship between the net injections at the nodes of a large transmission network and the flows over its elements, having made certain simplifying assumptions. In Section 4.2.2 we discuss how binding network constraints affect the incremental cost of electricity at different locations.

4.2.1 Power Flows

A transmission network can be considered as a set of power lines connecting nodes. Each node represents a substation, where generators and loads are connected to the network. A net injection at a node, at a given time, is the difference between the total production of all generators and total consumption of all loads located at this node taking place at that time. Power flows over each line of the network are determined by the net injections at all nodes as well as the network topology. The network topology is defined by the electric properties of power lines and the way they connect different network nodes.[1]

In power networks featuring alternating current (AC), the relationships between net injections and power flows over the transmission lines are not linear. However, in this chapter we use a linear approximation of these relationships, which is often called the 'DC (direct current) approximation'. We also abstract from transmission losses in the network. Such a model represents a realistic approximation that is often used to analyse the effects of loop flow and network interactions.[2]

Under the DC approximation, neglecting transmission losses, the power flow on each network element brought about by a set of net injections is determined by the following properties:

- *Superposition*   Power flows produced by a given set of net injections can be computed as the sum of flows produced by any combination of balanced net injections in which the initial set can be broken down.
- *Least resistance path*   The share of power that flows along each network path between each source node and sink node is inversely

Table 4.1   Net injections (MW)

Node	Net injection
1	6,000
2	1,500
3	−7,500

proportional to the relative resistance of such a path: the lower the resistance of a path the larger the share of power that flows along that path.

Table 4.1 and Figure 4.1 illustrate these properties and show how they can be used to compute the power flows brought about by a given set of injections and withdrawals. We considered a simple triangular network with three nodes connected by three identical lines. Such a network is the simplest example of a network with parallel paths and can be used to illustrate most of the network interactions taking place in real networks. We have assumed that generators are located at nodes 1 and 2 and that a consumption centre is connected to the transmission network at node 3.

We assume that transactions carried out on the market led to the (net) schedules of (net) injections shown in Table 4.1, and Figure 4.1 shows the setting of our example. Note that positive net injections offset the negative net injections, so that the total schedules over the network are

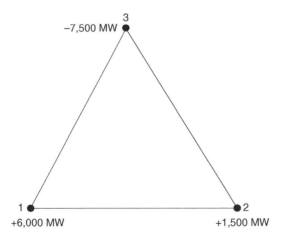

Figure 4.1   Triangular network example

balanced. Although we have assumed for reasons of simplicity that only generators are connected at nodes 1 and 2 and only loads are connected at node 3, this representation is general. Net injections at a node are the difference between total generation and total consumption taking place at the node. Therefore, a positive net injection indicates that generation at the node exceeds consumption, while a negative net injection indicates that consumption exceeds generation.

The set of net injections can be broken down into pairs of balanced injections and withdrawals, such that at each node the sum of net injections across all the pairs equals the total net injection. Thus, based on the superposition principle, power flows on the network will be determined as the sum of flows that would be independently caused by pairs of balanced net injections.

An example of the breakdown into balanced pairs of net injections is:

- a positive net injection of 6,000 MW at node 1, matched by a negative net injection of 6,000 MW at node 3; and
- a positive net injection of 1,500 MW at node 2, matched by a negative net injection of 1,500 MW at node 3.

The inverse resistance rule allows a simple computation of the flows resulting from each pair of balanced net injections. For example, in the first pair the injection at node 1 is balanced by a withdrawal at node 3. There are two paths between these two nodes: the short path $1 \rightarrow 3$, consisting of line 1–3 and the long path $1 \rightarrow 2 \rightarrow 3$, consisting of lines 1–2 and 2–3. Since all lines are assumed to be identical, the resistance of each path depends only on its length, and thus the resistance of the short path is half the resistance of the long path. The inverse resistance rule tells us that the flow produced by a transaction involving injections and withdrawals, respectively, at nodes 1 and 3 will split in the proportion of 2:1 between the short path and the long path. The flow on line 1–3 of the short path will be $2/3*6,000 = 4,000$ MW, and the flow over each of the lines 1–2 and 2–3 that make up the long path will be $1/3*6,000 = 2,000$ MW. This is illustrated in the left panel of Figure 4.2.

Applying the same logic, the second injection at node 2 balanced with the withdrawal at node 3 results in a flow of $2/3*1,500 = 1,000$ MW over line 2–3 and a flow of $1/3*1,500 = 500$ MW over each of the lines 2–1 and 1–3 that make up the long path between nodes 2 and 3. This is illustrated in the second panel of Figure 4.3. Note that the direction of flows over each line is shown by the sign relative to the direction of the

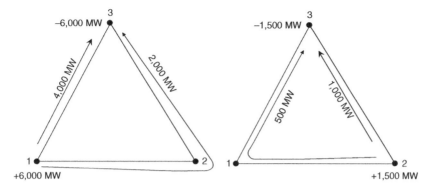

Figure 4.2   Power flows induced by injection-withdrawal pairs on a triangular network

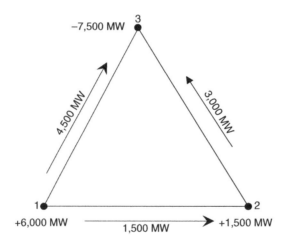

Figure 4.3   Superposition of the power flows

line: the flow of 500 MW over line 2–1 is equivalent to the –500 MW flow over line 1–2.

The superposition principle tells us that the total flows over each line of the network are the sum of the flows independently determined for each pair of balanced net injections, taken with the correct sign. The result is shown in Figure 4.3.

The principle of superposition in the DC approximation makes it simple to calculate the flows brought about by a set of net injections even for large networks, reducing them to a set of linear equations. This calculation first requires a systematic way to break down the initial set of net injections

Table 4.2   Net injections relative to the reference node 3 (MW)

Node	Net injection
1	6,000
2	1,500

into simple pairs of injections and withdrawals. This is done by selecting one of the network nodes as a reference or a swing node and assuming net injections at all other nodes to be balanced against that reference node. The initial set of net injections is then presented as a collection of independent net injections at all the nodes, each balanced against the reference node.

In the example above the two pairs of injections and withdrawals that we have considered were balanced at node 3, which means that we have chosen node 3 as the reference or 'swing' node. The choice of swing node is arbitrary and does not change the result of the computation. What is important is that the choice remains consistent throughout the calculation. The net injections relative to the reference node used in our example are shown in Table 4.2.

The power injected at each node and withdrawn at the reference node is split across the lines making up the parallel paths as indicated in the power transfer distribution factor (PTDF) matrix. The PTDFs are defined in relation to the reference node and depend on the technical features and topology of the network. The PTDF matrix used in our example relative to reference node 3 is shown in Table 4.3. Note that the positive or negative sign of PTDFs determines the direction of flows with respect to the line. The negative PTDF of 1/3 of node 2 relative to line 1–2 means that the net injection of 1 MWh at node 2 withdrawn from the reference node 3 generates a flow of 1/3 MWh over line 1–2 from node 2 towards node 1,

Once the reference node has been set and the PTDF matrix for the reference node has been computed, the DC power flows over all the transmission elements are calculated using a set of linear equations based on the

Table 4.3   PTDFs Matrix

Line	Node 1	Node 2
1–2	1/3	–1/3
2–3	1/3	2/3
1–3	2/3	1/3

net injections in all the remaining nodes. Power flow F_j over transmission element j is determined by:

$$F_j = \sum_i PTDF_{ij}^r \cdot I_i,$$

where, I_i is net injections at the node i balanced at the reference node r, and $PTDF_{ij}^r$ is the PTDF of the net injection at node i with respect to transmission element j, relative to reference node r. That is, the flow over each transmission element is the sum of net injections at all nodes of the network weighted by the flow sensitivities of this element to each node, given by the PTDF matrix.

4.2.2 Congestion and Locational Differentiation of Electricity's Incremental Cost

Network constraints
Elements of the transmission network may have certain limits on the power flow they can take. The simplest example is an overhead line that may overheat and stretch if the power flow is too large, with a risk of a short circuit. Transmission system operators often commit to maintain an N–1 reliability standard. This implies that the system should withstand an outage of any single transmission element at any time.

The N–1 standard results in contingency constraints. A contingency constraint limits the power flow over one line if another line, or network element, goes out of service. Since the outage may change the network topology, the contingency constraint may have different PTDFs compared with the all-in-service constraint on the same line. Regardless of the type, all constraints can be characterised by their PTDFs and the flow limit. Therefore, in the discussion that follows we do not distinguish between the types of constraint.

Security domain
Let us assume that in the setting introduced in the previous section, line 1–3 has a flow limit of 4,000 MW. In this case, the set of net injections assumed in the previous section becomes infeasible, since it induces a flow of 4,500 MW on line 1–3, beyond the line's capacity.[3] This is a congestion situation: generators and consumers wish to make injections and withdrawals producing flows that violate one or more network security constraints. In order to remain feasible, the net injections at nodes 1 and 2 have to be such that the power flows they create over line 1–3 remain within the line flow limits of 4,000 MW. Given the PTDFs of the net

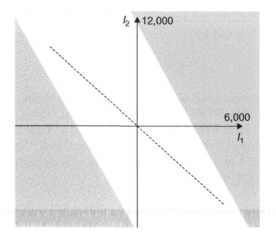

Figure 4.4   Security domain defined by the constraint on line 1–3

injections at nodes 1 and 2 with respect to this line of 2/3 and 1/3, respectively, this constraint can be written as:

$$\frac{2}{3}I_1 + \frac{1}{3}I_2 \leq 4{,}000$$

and the same constraint limiting the flow to 4,000 MW in the opposite direction is:

$$-\frac{2}{3}I_1 - \frac{1}{3}I_2 \leq 4{,}000.$$

These constraints identify the set of all possible combinations of injections and withdrawals that are feasible, that is, that produce flows that do not violate any constraint. We refer to this set of feasible injections and withdrawals as the 'security domain'. In the case of a network with three nodes, the security domain can be shown on a two-dimensional graph, for example in terms of net injections at nodes 1 and 2, assuming that all such net injections are balanced at node 3. The security domain defined by the constraint of 4,000 MW on line 1–3 is shown in Figure 4.4.

The security domain consists of the set of inequalities, each one limiting the flow along each line, expressed as a linear function of the injections at all nodes:

$$F_j = \sum_i PTDF_{ij}^r \cdot I_i \leq FL_j.$$

Here, F_j is the flow over transmission element j, I_i is net injections at the node i balanced at the reference node r, and $PTDF_{ij}^r$ is the PTDF of the net injection at node i with respect to transmission element j, relative to reference node r, and FL_j is the flow limit on constraint j.

Locational incremental cost

In the event of congestion the value of energy, that is, the minimum cost of meeting an incremental demand, differs at different nodes. Because of the network constraint, an additional withdrawal at a given node cannot necessarily be matched by increasing injections at the node where the cheapest available generation capacity is located. Instead, the lowest-cost option to match the additional demand at different nodes entails varying the injections of generators located at different nodes, and therefore results in additional costs.

In the setting introduced in the previous sections, let us assume that the incremental generation costs at nodes 1 and 3 are €20/MWh and €50/MWh, respectively, and that there is unlimited production capacity at both these nodes. At node 2 the generation cost is €10/MWh for production levels up to 1,500 MW, and very high for production levels above 1,500 MW. Note that these assumptions are consistent with the unconstrained market outcome assumed in the previous section: without network constraints the cheapest 7,500 MW generators in the system (1,500 MW at node 2 at €10/MWh and 6,000 MW at node 1 at €20/MWh) would sell their production on the market.

With the network constraint on line 1–3, the demand for 7,500 MW at node 3 is met at minimum cost in the following way. First, 1,500 MW is injected at node 2, as this node has the cheapest power. However, the next cheapest node, node 1, cannot meet the remaining 6,000 MW because the injection of 6,000 MW would breach the constraint on line 1–3. The remaining demand is therefore met by 5,250 MW injected at node 1 and 750 MW produced at node 3, giving a net injection at node 3 of –6,750 MW, in other words, a net withdrawal of 6,750 MW. This least-cost combination of net injections at nodes 1, 2 and 3 which respect the security constraint is shown in Figure 4.5.

We are now able to determine the lowest cost of matching an incremental demand increase at each node, that is, the value of energy at each node.

At node 1, an incremental withdrawal can be matched at minimum cost by increasing injections at the same node (production at node 2 is cheaper, but there is no spare capacity). Matching an incremental withdrawal with an injection at the same node does not change the net injection at the node, and therefore there is no impact on flows over the network. The

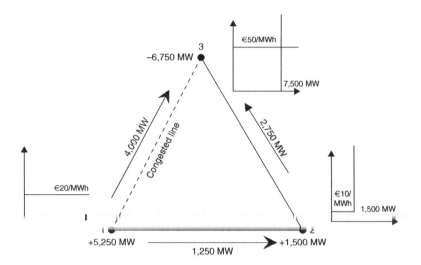

Figure 4.5   Power flows of the triangular network with congestion

incremental cost of withdrawals at node 1 is therefore €20/MWh, the marginal generation cost at node 1.

At node 3, increasing injections at the same node is also the least-cost way to meet an incremental load at this node. Although production at nodes 1 and 2 is cheaper, they cannot be used to meet additional demand at node 3. Additional production at node 1 would violate constraint 1–3 and at node 2 there is no spare capacity. The incremental cost of withdrawals at node 1 is therefore €50/MWh, the marginal generation cost at node 3.

The incremental cost of energy at nodes 1 and 3 is determined by the cost of the marginal units located at these nodes.

However, this does not hold for node 2, because of the physics of power flows. The cheap production capacity at this node is fully utilised to meet demand. Thus, additional demand at node 2 must be met by production at other nodes, such as 1 and 3. The cheapest way to match such incremental load without violating the constraint on line 1–3 is by increasing injections at nodes 1 and 3 by 1/2 MW each. In order to understand that this is the least-cost combination of injections matching the incremental withdrawal at node 1, consider that:

● the cost of this solution is 1/2*20 + 1/2*50 = €35/MWh;
● this solution creates zero incremental flow over the constrained line 1–3; and

- any other feasible combination of net injections at nodes 1 and 3 would require a larger share of production to take place at the more expensive node 3.

Thus, the incremental cost at node 2 is €35/MWh. This is determined by the cost of the variations in injections from the marginal generating units at nodes 1 and 3, so that these variations together produce zero incremental flow over the constrained line 1–3 and the total of these variations is 1 MW.

In the above example, the constraint on line 1–3 has restricted the power flows between nodes 1 and 3. As a result, the incremental costs of matching load in these nodes were set separately by marginal units at each of these nodes. Similar situations could arise if the power flows between nodes 1 and 3 were limited by constraints on lines 1–2 or 2–3. However, each of these situations would imply a different marginal cost of energy at node 2. Below we consider these two extensions of the previous example.

Extension 1: constraint on line 1–2

In this case we have assumed that line 1–2 has a flow limit of 1,000 MW, while there are no flow limits on other lines. This suggests that the minimum-cost injections and withdrawals shown in Figure 4.3 are not feasible. As in the main example, the cheapest capacity at node 2 is fully utilised and provides 1,500 MW. The remaining 6,000 MW of demand at node 3 is met by the generators at nodes 1 and 3. The least-cost way of doing this without violating the constraint on line 1–2 is to produce 4,500 MW at node 1 and 1,500 MW at node 3. This least-cost combination of net injections at nodes 1, 2 and 3, which respects the security constraint on line 1–2, is shown in Figure 4.6. Similarly to the main example, the incremental cost of energy at nodes 1 and 3 is determined by the cost of marginal units located at these nodes, that is, €20/MWh at node 1 and €50/MWh at node 3.

We shall focus on the incremental energy cost at node 2. This cost is determined by the least-cost variation of injections at nodes 1 and 3 that together make up 1 MW without changing the flow on the constrained line. A 1 MW injection variation at node 1 would have a 2/3 MW impact on the congested line, and a 1 MW variation at node 3 would have 1/3 MW impact on the congested line. Therefore, a double variation that would not violate the constraint would be a 2 MW increase to the net injection at node 3 and a 1 MW decrease in the net injection at node 1. The cost of such a variation would be $2*50 - 1*20 = €80/MWh$.

The marginal cost of energy at node 2 is higher than the cost of the marginal units at node 1 and node 3. This is because additional demand

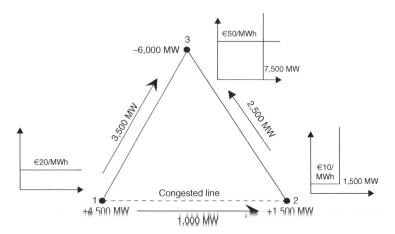

*Figure 4.6   Power flows of the triangular network with congestion:
constraint on line 1–2*

at node 2 increases congestion. To meet the demand without violating the
constraint, a costly counter-flow action would need to be implemented
between the other two nodes. Thus, the energy cost at this node reflects the
impact of this node on congestion in the rest of the network.

Extension 2: constraint on line 2–3

In this extension we have assumed that line 2–3 has a flow limit of
2,000 MW, while there are no flow limits on other lines. Again, this makes
the minimum-cost injections and withdrawals shown in Figure 4.3 infeasible. Contrary to the previous examples, it is no longer optimal to use the
cheapest capacity at node 2. This is because when node 2 is used together
with node 1 to meet the demand at node 3, line 2–3 very quickly becomes
congested and a large share of the load at node 3 needs to be met by the
expensive generators at that same node. It is cheaper not to use the capacity at node 2 at all, and to meet a larger share of the demand from the
generators at node 1. The resulting least-cost combination of net injections
which respect the security constraint on line 2–3 is shown in Figure 4.7. As
in the previous examples, the incremental cost of energy at nodes 1 and 3
is determined by the cost of marginal units located at these nodes, that is,
€20/MWh at node 1 and €50/MWh at node 3.

There is spare cheap generating capacity at node 2 at a cost of €10/
MWh. However, it is not this capacity that determines the incremental
energy cost at this node. Rather, as before, the energy cost at node 2 is
determined by the least-cost feasible variation of injections at nodes 1
and 3. This variation is a 2 MW increase of the net injection at node 1

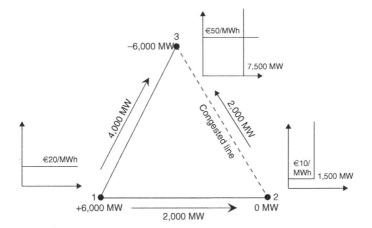

*Figure 4.7   Power flows of the triangular network with congestion:
constraint on line 2–3*

and a 1 MW decrease of the net injection at node 3. The cost of such a variation would be $2*20 - 1*50 = -€10/MWh$.

The marginal cost of energy at node 2 is negative. This is because additional demand at node 2 relieves congestion on the network. Meeting incremental demand at this node provides savings on the cost of the counter-flow action between the other two nodes.

General formulation of nodal incremental costs

In general, incremental energy costs at every node of the network are determined by an optimisation programme that minimises the cost of meeting demand in all locations by finding a set of generator injections at all network nodes, while ensuring that the flow constraints over all network elements are satisfied, in other words that the injections remain within the security domain. This optimisation programme is called 'security-constrained optimal dispatch'. A general outline of the optimisation programme can be written as:

$$\min_{I_i} \sum_i C_i(I_i).$$

The programme chooses the net injections at each node *i* that will minimise their total generation cost. Alternatively, it can be presented as a choice of net injections that maximise the total benefit:

$$\max_{I_i} \sum_i B_i(I_i).$$

Optimisation needs to ensure that the sum of the net injections over the entire system is zero. The optimisation therefore requires a balance constraint:

$$\sum_i I_i = 0.$$

Finally, optimisation requires a network constraint, in order to ensure that the chosen set of net injections remains within the security domain, or that the net flow over each line j stays below the line's capacity:

$$\sum_{i \neq r} PTDF_{ij}^r \cdot I_i \leq FL_j.$$

Incremental energy costs at all nodes are a byproduct of such dispatch optimisation. In the above examples the security-constrained optimal dispatch problems were simple enough to be solved manually and for the prices to be found. In large networks, security-constrained optimal dispatch becomes a complex optimisation problem. However, prices generated by these problems are relatively easy to analyse and to interpret thanks to several high-level properties. The following are some examples of these properties:

- Security-constrained economic dispatch determines the marginal cost of all flow constraints, sometimes called 'constraint shadow prices'. These represent the decrease in the total cost of meeting demand associated with a 1 MW increase in the constraint limit. The constraint shadow price is only non-zero if the constraint is binding in the constrained economic dispatch solution.
- At each node the difference between the incremental energy cost and the reference node energy cost is given as the sum of the constraint shadow prices weighted by the relevant PTDFs.
- In the security-constrained economic dispatch solution the number of marginal units is at least equivalent to the number of binding constraints plus one. The marginal units determine the incremental energy cost at their respective nodes. Prices at the remaining nodes are related to the cost of the marginal units via PTDFs.

In real networks with hundreds of nodes, no more than a dozen of constraints are usually binding at one time. Thus the majority of the nodal incremental costs are not independently set by the cost of the marginal units located at the respective nodes, but are interrelated via the PTDFs and shadow prices. This relationship is as follows:

$$P_i = P_r - \sum_j PTDF_{ij}^r \cdot SP_j.$$

In this formula, P_i is the incremental energy cost at node i, P_r is the incremental energy cost at the reference node, and $PTDF_{ij}^r$ is the PTDF of the net injection at node i with respect to transmission element j, relative to reference node r. Finally, SP_j is the shadow price of constraint j.

4.3 CONGESTION MANAGEMENT WITH NODAL PRICE DIFFERENTIATION

Congestion management is a collection of market (and sometimes non-market) arrangements that ensure that power flows produced by power injections and withdrawals do not violate the constraints, that is, that power injections lie within the security domain.

In this section we discuss congestion management methodologies that allow market participants to conclude only market transactions resulting in feasible power flows. This is achieved by allocating and enforcing a set of rights to inject and withdraw power at different network locations.

These rights to use the transmission resources on the electricity spot market can be defined and allocated in different ways. In Sections 4.3.1 and 4.3.2 we analyse methodologies in which the set of rights made available to the market reflects each and every network constraint. In one case these rights are defined explicitly as the rights to cause a power flow over individual critical elements of the transmission networks, or flowgates. In the other case the rights are defined implicitly by means of centralised clearing of the market at different electricity market prices at different nodes. In Section 4.3.3 we discuss long-term financial transmission rights that are used in systems where congestion on the spot market is dealt with using centralised nodal markets with locational prices.

In this section we consider locations to be individual network nodes, but in Section 4.5 below we discuss methodologies in which similar explicit or implicit arrangements are defined over a simplified representation of the network that groups multiple nodes into large zones.

4.3.1 Explicit Nodal Market: Flowgate Methodology

We have begun our analysis using the flowgate approach because it is based on a highly intuitive definition of transmission rights. In the flowgate approach, transmission rights are defined as the rights to induce a power flow through a given network element, or flowgate. This implies that a party willing to complete a transaction entailing power injection at

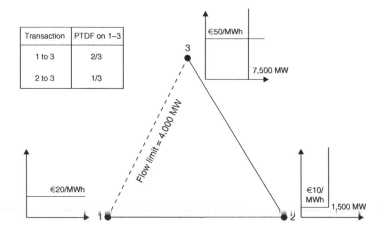

Figure 4.8   Triangular network with congested line and flowgate coefficients

a certain node and withdrawal at another node needs to procure the rights to use all network elements equivalent to the amount of the power flows over these elements[4] produced by the transaction.

For each network element the system operator issues a number of flowgate rights equivalent to the network element flow limit.[5] The system operator publishes the PTDFs of each flowgate relative to injections at each node matched with a withdrawal from the reference or swing node. These PTDFs can then be used to assess coefficients of flowgate utilisation by transactions involving any node pair.

We have used the simple setting developed in the previous sections to describe the market outcome in a system with flowgate rights. Figure 4.8 illustrates the example, which is consistent with those presented in Section 4.2.2 above. In this example the only network element with a flow limit is the line 1–3. The transmission system operator (TSO) issues the flowgate rights for this line for its flow limit of 4,000 MW and reports the PTDF coefficients representing usage of this flowgate by each transaction, specifically for the transactions from nodes 1 and 2 to node 3.

We have assumed that the generators at nodes 1, 2 and 3 carry out competitive bilateral trades to serve the demand of 7,500 MW at node 3. The value of the competitive bilateral deals over the network creates demand for use of the flowgate 1–3. The market price of these flowgate rights is determined by market participants' willingness to pay for the marginal MW of flowgate capacity.

For example, a generator at node 2 with a cost of €10/MWh is willing

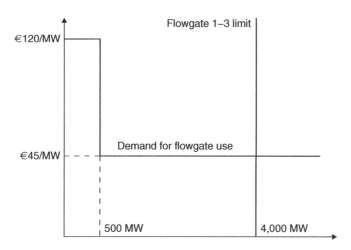

Figure 4.9   Demand for flowgate use

to pay up to €40/MWh for the opportunity to serve the load at node 3. If moving its production from node 2 to node 3 were more expensive, that generator would become uncompetitive compared with local production at node 3, which has a cost of €50/MWh and does not require flowing power through the flowgate. The PTDF matrix tells us that 1 MW injected at node 2 and withdrawn at node 3 generates a 1/3 MW flow through the flowgate. Therefore, the willingness to pay for 1 MW of the flowgate 1–3 by the generator at node 2 is 40/(1/3) = €120/MW. At that price, the entire available generating capacity of 1,500 MW located at node 2 would be activated to serve load at node 3, after purchasing 500 MW of the capacity of flowgate 1–3.

Likewise, a generator at node 1 with a cost of €20/MWh is willing to pay up to €30/MWh to serve load at node 3. This is the difference between the cost of local supply at node 3 and the cost of the generator at node 1. Since node 1 has 2/3 sensitivity on the flowgate 1–3, the generator is willing to pay up to 30/(2/3) = €45/MWh for each MW of the capacity of the flowgate 1–3. The large amount of capacity available at node 1 means that the entire capacity of the flowgate is sold at the price of €45/MWh to serve the load at node 3.

Figure 4.9 illustrates the demand and the supply for the capacity of flowgate 1–3. The clearing price of €45/MW for flowgate 1–3 is set at the intersection between the demand for flowgate use and the available flowgate capacity. The equilibrium of the flowgate and of the energy markets is such that the generator at node 2 produces 1,500 MW (impact of 500 MW on the flowgate) and the generator at node 1 produces 5,250 MW (impact of 3,500 MW on the flowgate). The remaining demand is met by generation located at node 3.

Note that this competitive solution in the system with flowgate rights coincides with the least-cost dispatch presented in the baseline example in Section 4.2.2. The competitive price of €45/MW of the flowgate constraint coincides with the shadow price of the constraint introduced in Section 4.2.2.

It is easy to verify that prices for electricity at each node can be calculated as the sum (or difference) of the marginal generation cost at the swing node (node 3 in our case) and the price of the flowgate right weighted by the sensitivity of the flowgate to the transaction between the given node and the swing node. In our example the price at the reference node 3 is set by the marginal cost of local generation of €50/MWh. The prices at nodes 1 and 2 are then calculated as follows:

$$P_1 = 50 - \frac{2}{3}45 = 50 - 30 = 20,$$

$$P_2 = 50 - \frac{1}{3}45 = 50 - 15 = 35.$$

More generally, the prices at all nodes in a market with flowgate rights are linked through the prices of all flowgate rights (SP) and node sensitivities with respect to those flowgates (PTDF):

$$P_i = P_r - \sum_j PTDF_{ij}^r \cdot SP_j.$$

Our simple example shows that the market equilibrium with congestion features differentiation of the electricity prices by location, and that these prices are directly related to the prices of the flowgate transmission rights. The prices that clear the energy and the flowgate markets satisfy a non-arbitrage condition: the electricity price at each node is equal to the electricity price at any other node increased by the transmission cost between the two nodes.

In the event that a flowgate becomes congested its clearing price is positive. The total value of this flowgate right, the product of the flowgate price and the available rights through that flowgate, is commonly referred to as 'congestion rent'. In the example above this value is 4,000 MW*€45/MW = €180,000. We discuss congestion rents extensively in the following sections.

In the framework of decentralised negotiation assumed in this section, market clearing is achieved through the usual bilateral negotiation process between buyers and sellers of electricity and flowgate rights. In Chapter 2 we discussed extensively the rationale behind centralising electricity trades taking place near time of delivery. We argued that transaction costs make it unlikely that the decentralised trading mechanism results in an efficient outcome at all times. The argument becomes even more compelling in the case of congestion, when the market-clearing prices for electricity and flowgate rights need to be discovered simultaneously. For this reason the flowgate approach has rarely been implemented in real electricity markets. Perhaps the only example is the electricity market of Texas, where flowgate rights were implemented in 2002 until the reform of the market in 2009–10.[6]

4.3.2 Implicit Nodal Market: Bid-based Security-constrained Economic Dispatch

In the previous section we discussed the impact of transmission congestion on the market outcome in a bilateral trading framework that highlights the analogies between electricity and all other goods. In this section we discuss the alternative market design in which electricity trading is centralised, and the rights to use the transmission network are allocated implicitly as part of the clearing process of the electricity market. The model is a stylised version of the standard market design of the nodal electricity market implemented in most US wholesale markets.

In this congestion management methodology the electricity spot market is cleared through an auction, like the one discussed in Chapter 2.[7] Generators and load-serving entities submit offers to sell and bid to buy electricity at the nodes where they are located. The clearing algorithm selects the set of offers and bids that:

- maximise the net surplus resulting from the transactions, and
- do not violate any network security constraints.

The market-clearing solution of security-constrained surplus maximisation produces a market-clearing price at each network node. This price is paid for all accepted sell offers and charged to all accepted buy bids at that node. In case of congestion the market-clearing prices at different nodes differ, to reflect the different cost of meeting the incremental demand at each location.

Flowgate rights are not explicitly allocated in the centralised setting; however, the set of transactions that market participants are allowed to carry out is limited to those meeting all the network-related constraints. For a given set of such constraints, with no transaction costs, the same market outcome would result from the centralised setting of a nodal market as from the decentralised setting with flowgate rights. In practice, however, because of transaction costs, the two implementations could yield different outcomes.

The market operator collects the congestion rent, which is the difference between the revenues from accepted bids to buy and the payments for the accepted offers to sell. It is simple to show that the congestion rent collected in this setting is identical to the congestion rent obtained in the alternative setting from the sale of flowgate rights. The congestion rent collected on the nodal market is the product of the net injections at each node and the clearing nodal price:

$$CR = -\sum_i P_i \cdot I_i = \sum_{i \neq r} (P_r - P_i) \cdot I_i.$$

Bearing in mind the relationship between the nodal prices and constraint shadow prices, this becomes:

$$CR = \sum_{i \neq r} \left(\sum_j PTDF_{ij}^{\,r} \cdot SP_j \right) \cdot I_i = \sum_j \left(\sum_{i \neq r} PTDF_{ij}^{\,r} \cdot I_i \right) \cdot SP_j.$$

If we use the relationship between the nodal net injections and flows over the transmission elements from the DC flow model, the congestion rent becomes:

$$CR = \sum_j F_j \cdot SP_j.$$

We should also recall that the constraint shadow price is non-zero only in the case of the constraint being binding and the flow being equivalent to the flow limit. Thus, the congestion rent is also equivalent to:

$$CR = \sum_j FL_j \cdot SP_j,$$

where FL_i is the flow limit transmission element j. This dual representation of congestion rent through nodal prices and nodal net injections on the one hand, and flow limits and shadow prices on the other, proves very useful when there is a need to identify the congestion rent generated by each transmission element.

4.3.3   Long-term Financial Transmission Rights

The spot market cleared at a nodal level may produce substantial differences between energy prices at different locations and create significant volatility in the congestion charges. Efficient trading may require instruments that allow market participants to hedge locational price risks. Forward financial transmission rights (FTRs), are instruments used on locational-based markets for that purpose.

An FTR with a contractual volume of 1 MW between injection point A and withdrawal point B entitles its holder to receive the difference between the locational spot prices at points B and A over the period covered by the contract. Therefore, if the volume of FTRs between two nodes matches the traded volume between them, an FTR becomes a perfect hedge against the cost of congestion.

FTRs typically have a duration of between one month and several years, and provide economic signals for the location of generators and customers. We discuss below how FTRs can also be used to provide incentives for investment in transmission.

Simultaneously feasible transmission rights and revenue adequacy

We assume here that the system operator, which also acts as the market operator, issues the FTRs. By issuing the FTRs the system operator commits to pay its holders the differences in energy market prices between pairs of nodes. This commitment is simple to implement when the net nodal injections resulting from the market clearing coincide exactly with the set of allocated FTRs. In this case the total FTR payment will precisely match the congestion rent collected by the market operator as a result of market clearing.

However, more often this is not the case and market-clearing injections and withdrawals differ significantly from the allocated FTRs. Yet this does not mean that the debts of the system operator to the FTR holders will outweigh congestion rents, placing financial risk on the market operator. In fact, provided the FTRs are simultaneously feasible, irrespective of the market outcome, the total payout to the FTR holders cannot exceed the congestion rent collected by the market operator. This result is commonly known as 'revenue adequacy'.

A set of FTRs is simultaneously feasible if the physical injections and withdrawals perfectly matching the set would not produce flows breaching any constraints. In other words, the set of FTRs is simultaneously feasible if the network can carry under security conditions the flows produced by their simultaneous exercise. In order to ensure that the set of FTRs allocated to the market is simultaneously feasible, the market operator carries out a simultaneous feasibility test. This process involves running an optimisation programme in which the objective function of the FTR allocation mechanism (for example, the auction value of the allocated FTRs) is maximised under the constraint that the sum of allocated FTRs at each node would produce feasible flows. This constraint can be presented as:

$$F_j^{FTR} = \sum_{i \neq r} PTDF_{ij}^r \cdot FTR_i \leq FL_j,$$

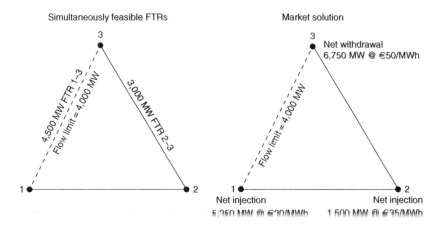

Figure 4.10   Simultaneously feasible FTRs

where F_j^{FTR} is the flow that would be produced by net injections matching the set of FTRs over line j, and FL_j is the flow limit of line j.

We shall illustrate revenue adequacy in Figure 4.10 using our usual example of a triangular network in which a single line 1–3 has a flow limit of 4,000 MW. We consider a set of FTRs comprising 4,500 MW FTRs between nodes 1 and 3 and 3,000 MW FTRs between nodes 2 and 3. It is simple to verify that this set of FTRs is feasible. The flow over line 1–3 produced by FTR 1–3 is 3,000 MW (2/3 of 4,500 MW) and the flow over line 1–3 produced by FTR 2–3 is 1,000 MW (1/3/ of 3,000 MW). Obviously, this is not the only possible set of feasible FTRs. For example, other feasible sets could be:

- 6,750 MW FTR 1–3 and 1,500 MW FTR 2–1; and
- 7,500 MW FTR 1–3 and 3,000 MW FTR 3–2.

Now let us assume that the market outcome was that of the previous example in Figure 4.5, above, that is, that the net injections at nodes 1, 2 and 3 were 5,250 MW, 1,500 MW and –6,750 MW respectively, and prices were €20/MWh, €35/MWh and €50/MWh, respectively. We can calculate the congestion rent accrued by the market operator as the sum of withdrawals multiplied by nodal prices less the sum of injections multiplied by the nodal prices:

$$CR = -5,250 \cdot 20 - 1,500 \cdot 35 + 6,750 \cdot 50 = €180,000.$$

The payout due to the holders of allocated FTRs can be calculated by multiplying each FTR by the difference between corresponding nodal prices:

$$FTR\,payout = (50 - 20) \cdot 4{,}500 + (50 - 35) \cdot 3{,}000 = €180{,}000.$$

This means that even though the FTRs do not initially match the market-clearing injections, the FTR payouts can be fully funded by the congestion rent collected on the market.

This is no coincidence. Readers can repeat this exercise for the other feasible FTR sets mentioned above. To illustrate this in a general sense we should recall that congestion rent can be expressed in terms of constraint shadow prices and the flow limits:

$$CR = \sum_j FL_j \cdot SP_j$$

The FTR payout, on the other hand, is:

$$FTR\,payout = \sum_{i \neq r} (P_r - P_i) \cdot FTR_i = \sum_j F_j^{FTR} \cdot SP_j,$$

where F_j^{FTR} are the flows produced by the set of FTRs on lines j. The revenue adequacy holds as long as the flows produced by the allocated FTRs over each line are within the limit, that is, as long as the FTR set is simultaneously feasible:

$$FTR\,payout = \sum_j F_j^{FTR} \cdot SP_j \leq \sum_j FL_j \cdot SP_j = CR.$$

We have shown that revenue adequacy holds in a simplified lossless DC approximation of transmission flow that we use throughout this book. However, revenue adequacy has been shown to hold in many flow models that are much closer to reality, such as a DC approximation with losses, or even an AC model. In fact revenue adequacy holds as long as the security domain defined by the network constraints is convex.

Most importantly, revenue adequacy holds as long as the network capacity remains the same from the moment the FTRs are allocated to the moment of market clearing. A change in network capacity could be, for example, a change in the flow limit of a constraint or the outage of a transmission element that could change both the PTDFs and some of the limits on the remaining transmission constraints. Even if the set of FTRs were feasible at the time of allocation, the changes in network topology could make this set infeasible on the network used for market clearing. Revenue inadequacy on a particular constraint will occur any time the constraint becomes binding in the market solution, and when the set of allocated FTRs induce a flow that exceeds the limit of this constraint.

Consider that in our example the flow limit of the line 1–3 was 4,000 MW when the simultaneous feasibility test was carried out, and therefore the awarded set of FTRs was feasible. Assume now that at the time of market clearing the limit decreased to 3,500 MW. The new limit does not change the market prices or constraint shadow price, but it does change the market-clearing volume of injections at nodes 1 and 3.

Since the prices are the same as before, the payout due to the FTR holders would also be the same, €180,000. However, we can calculate the new congestion rent, which will be the product of the constraint shadow price of €45/MWh and the new line limit of 3,500 MW, €157,500. We find that the decrease of the limit on the 1–3 line causes a revenue shortfall of €22,500 for the market operator.

Since FTRs are purely financial instruments, nothing prevents any economic agent from issuing them. However, as the discussion on simultaneous feasibility has shown, only the party that collects the congestion rent is hedged against the obligations falling on the FTR issuer. Any other issuer would be taking a purely speculative position on the price differences across locations. For this reason the system operator is the main if not the only issuer of FTRs in most electricity markets.

FTR allocation

There are many ways to allocate FTRs for existing transmission capacity. In most cases FTRs are auctioned off by the system operator to market participants. In FTR auctions the system operator may also select the set of FTRs that has the highest value for market participants, while being simultaneously feasible. The outcome of the auction is the allocation of FTRs to the market players and the setting of market-clearing FTR prices.

The FTRs achieve liquidity on the secondary markets, where market participants can exchange and reconfigure their FTR positions following developments in market conditions.

The revenue collected by system operators in FTR auctions is another form of congestion rent; it is actually the long-term expectation of congestion rent. These revenues are typically passed on to the transmission service customers, in the form of a reduction in transmission tariffs.

In some countries FTRs have been grandfathered to market participants based on pre-existing allocation of explicit flowgate-like transmission rights, or as a way of limiting the economic impact on existing generators of the introduction of a congestion management system based on locational prices.

FTR obligations and options

The discussion so far has assumed that FTRs are obligations, that is, the holder of an FTR is entitled to the difference between the market price at the withdrawal and injection point, regardless of whether the difference is negative or positive. If the price difference is negative, the holder pays the market operator the price difference.

In addition to these FTR obligations, market participants are often interested in the opportunity to hedge their position using FTR options. These options allow the FTR holder not to have to pay the market operator the price difference if it is negative.

An FTR option can be considered an FTR obligation that exists when the price difference between its withdrawal and injection points is positive, and disappears when the price difference turns negative. The feasibility of FTR options deserves special attention. A set of such FTRs should not only be feasible for injections and withdrawals matching all FTRs; it should also be feasible in case part of all the injections and withdrawals matching FTR options did not take place.

We can illustrate this in the context of the above example considering the following set of FTRs:

- A is an FTR obligation of 7,500 MW between nodes 1 and 3, and
- B is an FTR option of 3,000 MW between nodes 3 and 2.

The two FTRs are feasible if the price difference between nodes 3 and 2 is positive. The FTR A creates the flow over the constraint 1–3 of 5,000 MW, and the FTR B creates a counter-flow of 1,000 MW over this constraint. However, if the price difference is negative the flow matching FTR option B is not generated by the market equilibrium injections and withdrawals, making FTR A infeasible by itself.

The maximum number of FTR obligations between nodes 1 and 3 that can be allocated together with FTR option B is 6,000 MW. These FTRs will fully utilise network capacity in case the price difference between nodes 3 and 2 is negative so that FTR option B is not exercised. However, when the price difference is positive, the line capacity cannot be sold forward without placing financial risk on the market operator.

4.4 CONGESTION MANAGEMENT VIA RE-DISPATCH

Methodologies to deal with network constraints discussed in the previous section are based on restricting electricity market transactions to those that the network can safely accommodate. This is achieved by allocating to network users a feasible set of rights to inject and withdraw power at the different network locations. In the event of congestion, different electricity prices are set in different locations to enforce a feasible market outcome. These locational prices ensure that quantities that are economic to produce in each location induce flows over the network that do not violate transmission constraints.

In this section we discuss an alternative congestion management approach, consisting of two stages. In the first stage, market transactions are not subject to any network-related restrictions. Injections and withdrawals ensuing from this first stage may be infeasible, leading to flows that violate transmission constraints. This congestion is relieved in the subsequent stage, through re-dispatch, that is, by instructing and paying generators and possibly consumers to modify their injections and withdrawals at specific locations. In this section we analyse the features of the re-dispatch approach as well as certain bidding incentives that it creates. Later, in Section 4.6 we review the longstanding policy debate on the relative merits of congestion management via re-dispatch, nodal price differentiation or an intermediate zonal approach discussed in Section 4.5.

In Section 4.4.1 we introduce and illustrate the re-dispatch approach using a simple example. In Section 4.4.2 we present the bidding incentives created by this congestion management system and discuss some undesirable consequences of such bidding.

4.4.1 Principles of Re-dispatch Congestion Management

In the re-dispatch approach, market transactions taking place in the first stage are carried out assuming unlimited transmission capacity within national borders, and are not subject to any network-related restrictions. Market players buy and sell electricity, bilaterally or through power exchanges, knowing that the delivery and collection obligations corresponding to their sales and purchases can be fulfilled by, respectively, injecting or withdrawing power at any location of the network node of their choice. As a consequence, the energy market clears with a single system-wide price, as injections at any location are assumed to be perfect substitutes, as are withdrawals.

However, since no network-related restrictions are placed on the market players, power injections and withdrawals scheduled in the market equilibrium may violate network constraints. Congestion could be identified by the system operator based on generators' nominations (also known as 'schedules' or 'programmes') received after gate closure of the market.

Congestion resulting from the first unconstrained stage is relieved at a subsequent stage, through re-dispatch, that is, by paying generators and possibly consumers to reduce or increase injections and withdrawals at selected nodes.

The re-dispatch approach is widely adopted in Europe to relieve transmission congestion within national borders in such countries as the UK, France, the Netherlands, Spain and Germany. In part, the rationale for such market organisation is the perception that congestion within each national market is infrequent, and that little re-dispatch is required to make the unconstrained market solution feasible. We refer to such systems as re-dispatch or 'single-price' systems.

The second stage of re-dispatch congestion management is generally performed through a bidding market. This can be run immediately after the first-stage unconstrained market clearing, like in Spain, where the re-dispatch market (restrictions market) is run immediately after the unconstrained day-ahead single-price market. Alternatively, the re-dispatch market can be run closer to real time, as in the UK electricity market, where re-dispatch is performed as a part of the balancing mechanism.

Bids to reduce output collected on the re-dispatch market reflect the price that a generator is prepared to pay in order to scale back the production of a specific plant; conversely bids to increase output reflect the price that a generator agrees to receive in order to increase the production of a specific plant. In the event of congestion, the system operator selects the highest bids to reduce output compared with the day-ahead schedule from the generators located on the export side of the constraint, and the lowest bids to increase output from the generators located on the import side of the constraint. The aim of this bid selection is to modify the dispatch of plants in order to relieve congestion at the lowest cost while maintaining the energy balance within the control area.

Generators selected for schedule change during this process pay or are paid the value of their bids. Generators in the export area, whose bids to reduce output are accepted, buy energy back from the system operator at the bid price, instead of generating that energy themselves.

The price paid to the system operator by generators to reduce output in the export-constrained area is generally lower than the unconstrained market price. Conversely, the bid price to increase output paid by the TSO to generators in the import-constrained area is typically higher than the unconstrained market price. In this section we refer to the prices of accepted re-dispatch bids as 're-dispatch prices'.

As a result of re-dispatch, the system operator faces the costs of relieving congestion, since the cost of purchasing additional injections in the import-constrained area is greater than the revenues collected in the

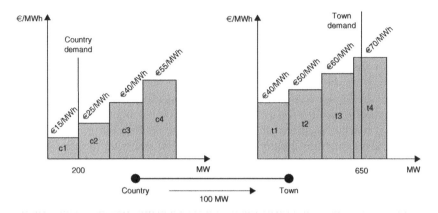

Figure 4.11   Example set up

export-constrained area from the generators called upon to reduce injections. These costs are generally socialised among market participants.

Example of re-dispatch congestion management
We illustrate the re-dispatch approach using a simple example. The focus of this example is bidding and bidding incentives rather than the effects of power flows over meshed transmission networks as in previous examples. Therefore we consider an example with two areas only, connected by a transmission link: Country and Town. The Country area has four generators, each with 200 MW of capacity. This area has consumption of 200 MW. The Town area also has four generators, each with 200 MW of capacity, but it has a higher energy demand of 650 MW. Generators located in the Country have somewhat lower variable production costs than generators located in the Town. The capacity of the transmission line connecting the two areas is 100 MW. This example is illustrated in Figure 4.11.

Now we assume that the demand in this market has been cleared in the day-ahead market. This market will be the first stage of the re-dispatch congestion management, in which the transmission constraints impose no limits on market transactions. To make it simpler we shall also assume for now that market participants are not aware of the presence of the transmission constraint.

We assume (as is quite often the case) that the day-ahead market has a non-discriminatory auction design. In this market the clearing price is determined by the highest accepted bid, and all accepted supply bids are paid this clearing price. In a competitive environment this auction design is known to induce bidding by generators close to marginal cost of

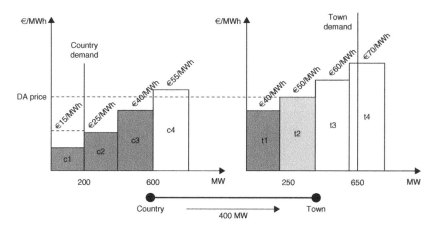

Figure 4.12   Unconstrained market outcome: no re-dispatch bidding incentives

production. The outcome of the day-ahead (DA) market, assuming such cost-based bidding takes place, is shown in Figure 4.12.

The least-cost way to meet the total demand of 850 MW in both areas is to use the cheapest available units in both areas. This involves scheduling at full capacity the three cheapest units located in the Country area, c1 to c3, and one unit t1 from the Town area. Finally, unit t2 is scheduled to produce 50 MW to meet the remaining 50 MW of demand. Unit t2 is marginal in the day-ahead market and sets the day-ahead price at €50/MWh. This day-ahead market outcome produces a power flow of 400 MW from the Country area to the Town area.

Upon clearing the day-ahead market the system operator realises that the power flow of 400 MW induced by the day-ahead schedules violates the constraint of 100 MW between the two areas, and re-dispatch is needed to reduce this flow by 300 MW. We consider first the cost-based merit order of units that are available to perform the necessary re-dispatch on both sides of the constraint. The least-cost way to eliminate the congestion involves reducing the output of the highest-cost scheduled units in the Country area by 300 MW. This requires reducing the entire output of c3 and 100 MW of the output of c2. The least-cost way to eliminate the congestion involves increasing the output of the lowest-cost available units in the Town zone by 300 MW. That requires increasing the output of t2 by 150 MW and increasing the output of t3 by 150 MW.

Unit c2 is the marginal unit running in the export-constrained Country area after this cost-based re-dispatch, and the variable cost of c2 (€25/MWh) represents the marginal cost of energy in this area. Unit

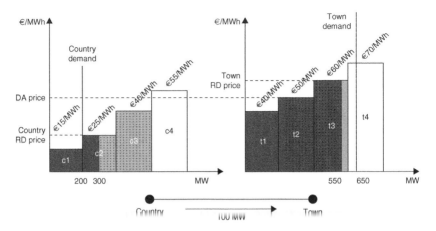

Figure 4.13   Cost-based re-dispatch and the equilibrium area prices

t3 becomes the marginal unit running in the import-constrained Town area in the cost-based re-dispatch, and its variable cost of €60/MWh is the marginal cost of energy in this area after taking into account the constraint. In a competitive re-dispatch market, these marginal energy costs in each area would determine the competitive prices in each area. Such competitive prices would also be set in the case of locational market clearing.

Figure 4.13 illustrates the re-dispatch (RD) prices determined by the marginal cost of the marginal unit running in each area after cost-based re-dispatch, as well as the day-ahead price that would prevail in the absence of congestion.

4.4.2   Bidding Incentives Created by Re-dispatch Congestion Management

The re-dispatch market may have an important impact on the bidding behaviour of competitive generators: if market participants can predict that there will be congestion that will be resolved in the re-dispatch market, they will bid differently in the day-ahead market.

Day-ahead bidding by competitive generators expecting a transmission constraint

When we first considered the unconstrained day-ahead market in our example, we assumed that market participants would be unaware about the network constraint, and each competitive generator would bid according to its marginal cost. However, in a market where congestion management is carried out via re-dispatch, if congestion is predicted, rational

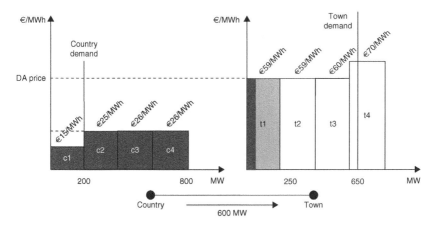

*Figure 4.14   Illustration of the arbitrage between day-ahead and
                re-dispatch market*

competitive generators will realise that submitting cost-based day-ahead
bids as presented above is not profit maximising.

Generators would expect that, because of the transmission constraint,
the re-dispatch price at their location would be different from the
nationwide uniform day-ahead energy price, providing them with an
opportunity to arbitrage between these two markets. In order to perform
arbitrage between the day-ahead and the re-dispatch market, generators
would submit bids in the day-ahead market departing from their variable
production costs. Instead, their day-ahead bids would be driven by the
opportunity costs determined by the price they could get in the re-dispatch
market in their respective areas.

In our example, these bidding incentives mean that the entire capacity of
unit c4 in the Country area and the entire capacity of unit t1 in the Town
area will be bid for in the day-ahead market at opportunity cost. The
opportunity costs are aligned with the competitive re-dispatch prices in
their respective areas. The outcomes of the day-ahead and the re-dispatch
markets are illustrated in Figures 4.14 and 4.15.

Figure 4.14 suggests that all generators in the Country with a variable
cost above the re-dispatch price would submit day-ahead bids at that
re-dispatch price. At the same time, all generators in the Town with a
variable cost below the re-dispatch price would submit day-ahead bids at
that re-dispatch price. The re-dispatch price in each area represents the
opportunity cost of these units.

Such bidding changes the outcomes of the day-ahead market compared with the cost-based bidding presented above. In the first place, the

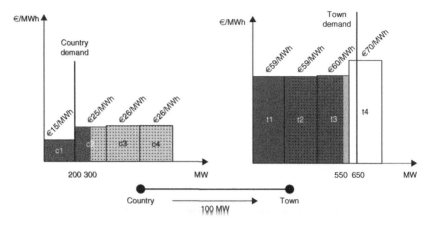

Figure 4.15   Outcome of the re-dispatch market

day-ahead price may change. In this example the marginal unit setting the day-ahead price becomes t1 with its bid of €59/MWh. In the second place, the day-ahead dispatch results in a flow from the Country to the Town of 600 MW, requiring a re-dispatch of 500 MW, which is 200 MW more than under the cost-based day-ahead bidding. This also increases the overall cost of re-dispatch as illustrated in Figure 4.15.

Three additional aspects of the incentives discussed above are worth mentioning: market power, auction design and congestion certainty:

- These bidding incentives arise in the case of perfect competition in each area of the market, and are not related to the exercise of market power. When exercising market power, generators bid differently from their variable production costs in order to modify the market price. When doing so they accept losing sales of a part of their capacity in order to receive increased profits from the remaining sales due to the higher prices. This market-power logic is not the one that drives the bidding incentives discussed above.
- The market outcome that we have discussed is to a large extent independent of the design of the day-ahead and re-dispatch markets. In the example we assumed that the re-dispatch market is run as a pay-as-bid auction. As discussed in Chapter 2, this means that the competitive generator in the import constrained area offers just below the highest price expected to ensure that an offer is accepted; similarly a competitive generator in the export constrained area forecasts the lowest price that would ensure that a bid is accepted and bid just above that price. In case the re-dispatch market were

run as a non-discriminatory auction the bidding strategy of the generators would be different because accepted bids and offers respectively pay and receive the market clearing price, irrespective of the bid and offered prices. However, changing the auction design does not remove the fundamental incentive to arbitrage between the day-ahead and the re-dispatch prices.

● In this example we have assumed that generators have perfect information about the market, which allows them to accurately predict the competitive re-dispatch price in each area. In reality there exists a certain degree of uncertainty about the market parameters, which makes such prediction difficult. As a result, the competitive bidding strategies described above may present a certain degree of risk for generators. For example, if there is a high probability that the expected congestion will not occur, such bidding could on average result in losses rather than being profitable. The more uncertainty there is, the closer the day-ahead bids of competitive generators are to their variable cost.

4.5　THE INTERFACE OR ZONAL APPROACH

In this section we discuss the approaches to congestion management that use a simplified representation of the network, where network nodes are grouped in large zones and the constraints on individual transmission elements are grouped into interface constraints between zones.

These approaches mix the two approaches previously discussed. On the one hand, similarly to the nodal approach, transactions in the electricity market are constrained by the flow limits over the simplified representation of the network. On the other, if transmission congestion occurs within the zones defined by the simplified network representation, it is resolved using the re-dispatch approaches. The zonal approach has traditionally been implemented in Europe to allocate cross-border transmission capacity, in which case each country is considered a market zone.

Just as in the case of the nodal model, the rights to use transmission capacity over the simplified zone transmission model in the spot market can be defined explicitly and implicitly. Explicit rights between the exporting and importing zones allow spot transactions to be scheduled with power injections anywhere in the exporting zone matching withdrawals anywhere in the importing zone. Implicit rights are allocated through centralised clearing of the spot-market bids, and allow different spot prices to be set in the market zones in case the transmission flows over the interface constraints reach their limits.

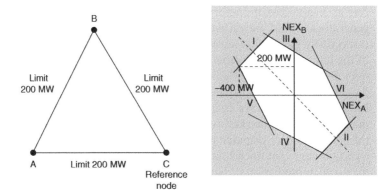

Figure 4.10   Transmission network and security domain in a 3-zone market

The approach used to simplify the real network into a zone representation is crucial in determining the amount of transmission capacity available between the market zones and for the resulting market outcomes. We discuss in the following sections two such approaches or capacity models: the net transfer capacity approach that has traditionally been used in Europe; and the flow-based approach that is currently being implemented in several regions of Europe.

4.5.1   Capacity Model in a Three-market Security Domain

In this and subsequent sections we discuss the zonal representation using a simple example of three countries: A, B and C, each country being a market zone. We first assume that the actual transmission network connecting the countries is rather simple (later in Section 4.5.3 we move away from this assumption) and similar to the triangular networks considered previously in this chapter. Each country is represented by a single node interconnected by transmission lines with identical impedance and an identical flow limit of 200 MW, as presented in the left panel of Figure 4.16.

First, we shall describe the characteristics of our network's security domain. The security domain is a combination of all possible sets of injections and withdrawals that do not violate any security constraints. In our three-market zone setting, each set of net withdrawals (conventionally called 'net exports' when referred to the zonal markets) from each market zone can be represented as a point in a two-dimensional space. The net exports in the third market zone chosen as the reference

Table 4.4 PTDFs of a triangular network

Line	NEX_A	NEX_B
AB	−1/3	1/3
BC	−1/3	−2/3
AC	−2/3	−1/3

market will be uniquely determined by the balance condition, that is, the condition that the sum of the net exports in the three market zones is zero.

The right panel of Figure 4.16 shows the security domain in terms of the net exports from markets A and B (NEX_A and NEX_B).

For this network, the PTDFs relative to the reference market C are shown in Table 4.4. As we discussed in the previous sections, the values of these PTDFs follow on from that fact that in this triangular network, for each MWh of transaction between two markets, the power flows split over parallel paths in the proportion of 2:1 according to the length of the parallel paths.

Each side of the security domain shown on Figure 4.16 represents the flow constraint on each transmission line. The constraint limiting the flow on line AB within 200 MW in either direction identifies the set of net exports so that:

$$-200 \leq \frac{1}{3}NEX_A - \frac{1}{3}NEX_B \leq 200.$$

This condition is satisfied by all combinations of NEX_A and NEX_B that lie in the region between lines I and II in the figure.

The flow constraint on line BC of 200 MW in both directions identifies the set of net exports so that:

$$-200 \leq \frac{1}{3}NEX_A + \frac{2}{3}NEX_B \leq 200.$$

This condition is satisfied by all points in the region between lines III and IV in the figure.

Finally, the constraint of 200 MW on line AC in both directions identifies the set of net exports such that:

$$-200 \leq \frac{2}{3}NEX_A + \frac{1}{3}NEX_B \leq 200.$$

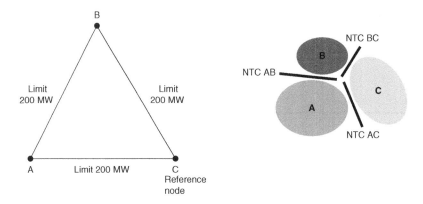

Figure 4.17   NTC transaction rights for a triangular network

This condition is satisfied by all net exports in the region between lines V and VI in the figure. The white surface in the figure above is the security domain of our network. Any set of net exports that lies within this area is feasible, since it meets all of the security constraints.

4.5.2   The Net-transfer-capacity Capacity Model

The traditional implementation of the interface methodology in Europe is commonly referred to as the 'net transfer capacity' (NTC) model. The main feature of the NTC model is that it sets limits on bilateral transactions between each pair of neighbouring countries regardless of the exchanges between the other countries. The NTC capacity model defines a matrix that determines the limits to bilateral trades between each pair of neighbouring markets (Figure 4.17).

At the initial stage of electricity market development in Europe, the system operators would issue and auction off (or otherwise allocate) explicit transmission rights corresponding to the NTCs on each border. These transmission rights would entitle the holders to schedule matching injections and withdrawals between the corresponding neighbouring countries.

However, this approach has often led to situations where available NTC capacity was not fully utilised in spot-market trades, or even utilised to schedule flows inconsistent with the spot-price difference, that is, implementing exports from the more expensive to the cheaper market. The efficiency of the cross-border transmission capacity utilisation was dramatically improved by the introduction of implicit auctions on some borders, where the allocation of cross-border NTC capacity is integrated with clearing of the spot markets.

Table 4.5   NTC matrix

From/to	A	B	C
A		50	200
B	100		150
C	250	200	

The NTC capacity model is one of the ways to set limits on feasible commercial transactions, that is, to define a transaction space within the security domain. Using our simple example we analysed the transaction space provided by the NTC capacity model compared with the security domain. We highlight three important features of the NTC model:

● the same physical network can support multiple alternative NTC transaction spaces;
● a given feasible NTC transaction space may not be able to cover the entire security domain. Achieving some feasible transactions within the NTC model might mean not being able to achieve others; and
● in some cases there may be no feasible NTC transaction space that would allow the market to reach certain parts of the security domain.

A set of NTC limits on bilateral transactions between pairs of neighbouring countries (regardless of the exchange between other countries) implies that net exports from each country are limited by a simple sum of NTC limits between this and all the neighbouring countries. For example, consider the matrix of NTC limits in Table 4.5. This matrix implies that the maximum NEX_A is given by the sum of NTC_{AB} and NTC_{AC} and is 250 MW, while the minimum NEX_A is given by the negative of the sum of NTC_{BA} and NTC_{CA} and is –350 MW.

Thus, the transaction space of the NTC capacity model can be presented in the form of individual limits to the net exports from each market. This space can be illustrated on a two-dimensional graph. Because the NTC capacity model sets individual limits, the zone net exports and the boundaries of the transaction space can only be represented by vertical lines, horizontal lines, or diagonals with an angle of –45°. For example, the maximum and minimum limits on the NEX_A are given by the vertical lines, the maximum and minimum limits on the NEX_B are given by the horizontal lines, and the maximum and minimum limits on the NEX_C are given by the diagonal lines. Along each of these diagonal lines the sum of NEX_A and NEX_B is constant and we know that

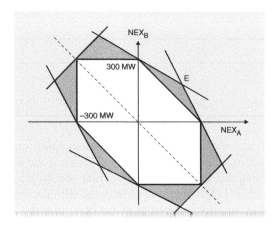

Figure 4.18   The largest NTC transaction space and the security domain

NEX_C is equal to the sum of NEX_A and NEX_B taken with a negative sign. With these vertical, horizontal and diagonal lines, the shape of the NTC transaction space on a two-dimensional graph looks like either a square or a diamond. To make sure that any transactions on the market produce feasible transmission flows, the transaction space must lie entirely within the security domain.

Figure 4.18 illustrates the largest transaction space that can be supported by the security domain in our example. It is easy to verify that a vector of NTC levels making up this transaction domain is 150 MW across each border in both directions. The figure shows that not all the feasible transactions can be achieved within this set of NTCs. For example, the transactions represented by point E in the figure correspond to 200 MW imports into market C from market A and 200 MW from market B. This set of transactions is compatible with the network's capacity, as it results in a flow of 200 MW on lines 13 and 23 and 0 MW on line 12. Nevertheless this set of transactions is not allowed by the selected NTC levels of 150 MW on each border.

Note that one could choose a set of NTC values so that point E is included in the transaction space, for example: 200 MW between zones A and C, 200 MW between zones B and C and 0 between A and B. The resulting transaction space is shown with a straight line in Figure 4.19. As shown in the figure, the inclusion of point E in the transaction space means excluding a large number of other feasible points belonging to the security domain. That means that when selecting the set of net transfer capacities allocated to the market, system operators must conjecture upfront which is the highest-value set of transactions. In our example, the system oper-

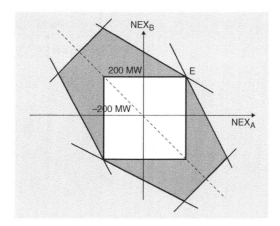

Figure 4.19   The NTC transaction space including point E

ators would select the NTC including point E in the transaction space if they knew that generation costs in market C are much higher than in both market A and B, and that generation costs in market A and B are similar.

In the previous example we showed that a feasible transaction space defined by a given set of NTC limits might cover only a part of the security domain. In this case one can only select a transaction space that includes a specific point of the security domain at the expense of other feasible points, which would be put out of the market's reach.

In addition, in some cases, there is no choice in the NTC limits allowing market transactions in certain portions of the security domain. This happens because certain points of the security domain are only feasible when a particular transaction creating a counter-flow is scheduled. Consider the network shown in Figure 4.20. The transaction space shown in this figure as the area delimited by the solid dashed lines corresponds to the following set of NTC limits: NTC_{BC} and NTC_{AC} are 600 MW, while NTC_{AB}, NTC_{BA}, NTC_{CB}, and NTC_{CA} are all zero. One can interpret this selection of net transfer capacity as aiming to maximise imports into (or exports from) Country C. According to the figure, a large part of the security domain does not belong to the transaction space. In particular, a number of transactions that the network can support and that would result in greater imports into C are not allowed. Yet it is impossible to expand the transaction space any further. Point E, for example, in which market A exports 900 MW towards market C, is feasible only if a counter-flow transaction from B to C of 300 MW is scheduled. However, in the NTC model there is no means of identifying such a condition since each transaction is considered independently from the others.

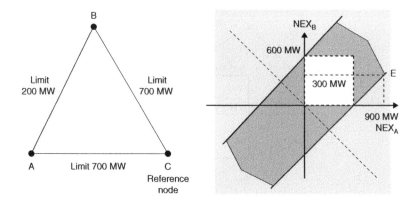

Figure 4.20   Transaction space of the explicit flow-based auction

Therefore, because of its simplicity the NTC model is unable to allow market participants to explore the entire security domain.

4.5.3   The Flow-based Capacity Model

The development of power exchanges in many European countries and increasing coordination among system operators have paved the way for the introduction of a new capacity model that is potentially more efficient than the traditional NTC approach. This new methodology is commonly referred to as 'flow based'.

In the flow-based approach, each constraint sets a limit to the flow over a critical infrastructure or critical branch, a physical transmission network element affected by the cross-border transactions. This constraint then sets a ceiling to the weighted sum of the net exports from all market zones, where the weights are the PTDFs of these market zones with respect to each critical branch. The constraints have a similar structure to those discussed in Section 4.3:

$$\sum_{i \neq r} PTDF_{ij}^r \cdot NEX_i \leq FL_j$$

Thus the set of flow-based constraints and the transaction space defined by the flow-based capacity model coincide with the security domain. In the example presented in Section 4.5.1, the flow-based transaction space would match the security domain shown in Figure 4.16.

This represents a major improvement on the NTC model where, as we saw above, the transaction space may cover only parts of the security domain. For this reason, use of the flow-based capacity model is expected to allow larger volumes of cross-border trade and a more efficient use of the existing physical network in general.

Currently two versions of flow-based methodology are being developed in Europe. One is flow-based market coupling between France, Belgium, the Netherlands, Germany and Luxembourg (the Central West Europe region, or CWE). In the CWE, the day-ahead electricity markets of the participating countries are simultaneously cleared in such a way that the trade surplus is maximised subject to flow-based constraints.[8] The rights to use the transmission system are thus implicitly allocated, as we discussed in Section 4.3.2.

The second version of flow-based methodology is being developed for the allocation of explicit short-term transmission rights between Austria, the Czech Republic, Germany, Hungary, Poland, the Slovak Republic and Slovenia (the Central-East Europe region, or CEE). In this model, the markets for energy and for transmission rights are not cleared simultaneously. First market participants submit bids for transmission rights between any pair of participating national markets. The auction for these rights is cleared in such a way that the value of the awarded transmission rights is maximised subject to the set of flow-based constraints,[9] similarly to the FTR allocation process in the nodal markets discussed in Section 4.3.3. The transmission rights allocated in this process give their holders the rights to schedule matching transactions between market zones in the day-ahead market.

The transaction space provided by the explicit version of the flow-based capacity model should be expected to be smaller than the one that the implicit model could achieve. This is because the transmission rights allocated in the explicit model have the character of an option. As we discussed in Section 4.3.3, such option rights may not be capable of fully utilising available transmission capacity or fully exploiting the counter-flows that can increase transmission capacity.

4.5.4   Accounting for Nodal Details in the Flow-based Capacity Model

The interface approach addresses network constraints by limiting net power transfers between large areas of the network, or zones. In the examples discussed thus far we have assumed that each market zone could realistically be approximated by a single network node. In other words we assumed that, power injections at any part of the market zone have the same impact on flows over all the critical branches.

In this section we address issues related to the actual size of the zones in the context of the interface approach. Specifically we consider situations where net injections at different nodes of the same zone could cause

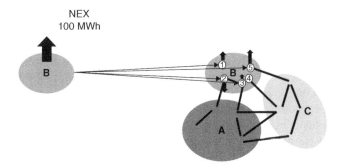

Figure 4.21   The link between the net exports and nodal injections through GSKs

different flows across critical infrastructures. This could happen in a wide range of situations, especially when a single market zone comprises a large country. This may happen, for example, when a critical infrastructure is located entirely within the zone. In this section, we refer to the most efficient version of the interface approach, flow-based market coupling. However, similar issues arise in all other implementations of the interface approach.

When different nodes in a zone have a different impact on a critical branch, the same value of the total net exports from the zone can have a different impact on the critical branch, depending on how the net exports are distributed among the nodes of that zone.

In Section 4.3, above, we noted that the impact of each node on the flow over a constraint is given by a PTDF. In the flow-based approach, such node-level PTDFs are synthesised into zone-level PTDFs taking weighted averages of the node-level PTDFs over all nodes belonging to a given zone. The nodal weight factors used in this process are called 'generation shift keys' (GSKs).[10]

The GSK of each node is the share of the zone's net injections that is expected to take place in that node in real time. Figure 4.21 illustrates the link between the net exports from the zone and the corresponding nodal injections through GSKs. The zone-level PTDF for a critical branch is then calculated by summing the PTDFs for that branch of all nodes belonging to the zone, weighted by the GSKs:

$$PTDF_{zj} = \sum_{i \in z} PTDF_{ij} \cdot GSK_i^z,$$

$$\sum_{i \in z} GSK_i^z = 1.$$

If the actual distribution of injections across the nodes grouped in the zone is different from the expected distribution reflected in the GSKs, the actual flows over the critical branches will be different from the flows implied by zone-level PTDFs. The same level of net exports from the market zone can induce an unexpected flow over the critical branches if the actual distribution of nodal net injections diverges from the one reflected in the GSKs. Specifically, the actual flows over a critical infrastructure may turn out to exceed the security limit even if the net injections meet the constraints defined at the zone level. This possibility is dealt with by reducing the capacity of the critical infrastructure made available to the market by the flow reliability margin (FRM).[11] The system operators assess the variation in the flow over each critical branch resulting from errors in the GSK prediction and use the extreme values of this variation to set the FRM for the branch.

Application of the FRM reduces the transaction space compared with the security domain achievable under the nodal approach. The FRM and reduction of the trading space are all the more significant for large national markets in which transmission nodes in different parts of the country have very different impacts on the flows over the cross-border critical branches.[12]

The choice of GSKs may also have an important impact on the market outcome in the flow-based approach. In Section 4.3 it was shown that the locational market-clearing price of electricity may reflect the marginal cost of congestion that incremental demand in this location creates or resolves in other parts of the network. The GSKs may have an impact on the valuation of that impact on congestion, often providing a biased valuation.

We illustrate this effect in the usual three-zone market. The setting of our example is illustrated in Figure 4.22. We assume for simplicity that in

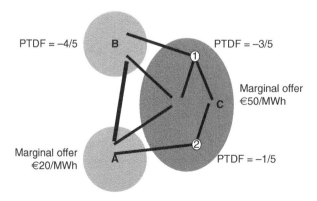

Figure 4.22   Impact of GSKs on market prices

zones A and C unlimited generation capacity is available at variable costs of €20/MWh and €50/MWh, respectively, and that in equilibrium the link between market zones A and B is constrained, while all remaining critical branches have ample capacity. In this example we also assume that zone C consists of two nodes that have a very different impact on the binding constraint. The PTDF of node 1 calculated relative to the reference market A is –3/5 (close to the PTDF of market B, –4/5), while node 2 has a PTDF of –1/5 (close to the PTDF of market A, 0). In order to isolate the effect on the market outcome of the methodology to set the GSKs, we assume that the market outcome can be perfectly predicted by the system operator. This rules out any distortions related to prediction errors.

Under our assumptions, the market equilibrium prices in zones A and C are equal to the marginal generation costs, €20/MWh and €50/MWh, respectively. Given the zone level PTDFs, the market clearing price in zone B can then be determined through the relationship, discussed in Section 4.3:

$$P_z = P_r - \sum_j PTDF_{zj}^r \cdot SP_j.$$

For each selection of GSKs one set of zone-level PTDFs and therefore one price level in B are determined. We consider three cases of GSKs, one that puts 100 per cent of the weight on node 1, one that puts 100 per cent of the weight on Node 2, and one that weights the two nodes equally. Table 4.6 shows calculation of the prices in market B for these GSK cases.

The zonal PTDFs corresponding to the three GSK scenarios are 3/5, 1/5 and 2/5, respectively. The constraint shadow price can be calculated for each value of zonal PTDF as the difference between the prices set by generators in zones C and A, €30/MWh, divided by the PTDF of zone C. Finally, the price in market B in each scenario is calculated using the above formula, with the prices in reference zone A, the constraint shadow price, and the PTDF of zone B.

The table shows that the price in B is highly dependent on the choice of GSKs. Prices would reflect the correct marginal cost of congestion created

Table 4.6 GSK impact on market prices

GSK	GSK weights		PTDF of zone C	Constraint shadow price	Price market B
	Node 1	Node 2			
1	1.0	0	–3/5	50	60
2	0	1.0	–1/5	150	140
3	0.5	0.5	–2/5	75	80

by the demand in each zone if, in each zone, the node where the marginal plant is located receives GSK weight 1. Such marginal GSK weighting is very difficult to implement, since it is much more difficult to predict the marginal plant than the average contribution of each node to the zone's production. However, in most cases the average GSK proportional to actual production in each node will produce zonal prices incorrectly reflecting the marginal cost of congestion. This may result in inefficient distribution of production across the zones. In our example, the market equilibrium production level in B could be materially higher if the clearing price was €140/MWh rather than if the price was €60/MWh.

Furthermore, even though GSKs related to the marginal units could provide the efficient price outcome, such GSKs may result in incorrect assessment of the flows produced by the net exports from each market zone over the critical branches. These flows depend on the net injections in all nodes, and not only on the injections at the marginal node. Only average GSKs that are proportional to the actual nodal production and the actual nodal share of the total zonal export correctly predict the flows on the critical branches that will result in market equilibrium. Using marginal rather than average GSKs may then require larger FRM, to the detriment of the transaction space.

Perfect GSKs, which would both lead to efficient prices and allow correct assessment of the power flows over critical branches, do not exist.

4.6 THE LOCATIONAL PRICE AND RE-DISPATCH DEBATE

In this section we present some elements of the longstanding debate about the pros and cons of the approaches to congestion management discussed in Sections 4.3 and 4.4, above: the nodal and the re-dispatch approaches. We also refer to current European developments in zonal approaches that to some extent mix the elements of the nodal approach, in the flow-based market coupling of market zones, and the re-dispatch approach to congestion within the zones. The debate is ongoing in the context of European electricity market integration, particularly with respect to of the optimal size of the market zones in flow-based market coupling.

We discuss the following aspects of the debate: the cost of congestion, market power, market liquidity, implementation and operation complexity, incentives for generator investments, incentives for network development, political pressure for a single national price for customers, and the wealth distribution properties of the two approaches.

4.6.1   Cost of Congestion

Nodal congestion management has been argued to result in a higher cost of congestion eventually paid by market participants than the alternative single price system with re-dispatch. Several formulations of this critique have been brought forward.[13] These critiques can be illustrated by a simple example, where a generator with a very high cost located in an import-constrained area has to be turned on to produce a small amount of energy in order to relieve congestion. In the nodal approach, this generator considerably increases the clearing price that is paid by all customers and received by all other generators in this location. It is argued that if the same congestion was dealt with using the re-dispatch approach, the impact on the market would be much smaller, since the cost of re-dispatch would amount only to the cost of the additional power from this generator.

This reasoning has several flaws. First, it neglects the congestion rent that is collected by the system operator in the nodal approach when there is congestion, which is eventually returned to market participants. Part of the increase in the locational price in the import-congested area will therefore be hedged by the congestion rent. The total cost of congestion to customers after the congestion rent is passed on to them may be similar to the cost in a re-dispatch approach.

Second, this example assumes that in the re-dispatch system competitive generators would offer their capacity at variable cost. However, this is not the case, as discussed in Section 4.4. Because of the day-ahead bidding at opportunity cost induced by re-dispatch congestion management, customers could eventually face a higher cost of congestion than in the nodal approach. On the other hand, the nodal market clearing provides an incentive to competitive generators to offer their power at marginal cost and does not induce any opportunity cost bidding.

A more substantial argument on the cost of congestion relates to markets where transmission congestion has to be addressed by the re-dispatch approach in time, through balancing market transactions. Performing such real-time re-dispatch may turn out to be unnecessarily expensive for the system operator, because in real time it may have a limited choice of fast and flexible generating resources available to provide re-dispatch. Congestion could be resolved more cheaply if it were addressed at an earlier stage, for example, in the day-ahead market, where there is enough time to start up and ramp up inflexible units that could deal with congestion most efficiently.

As the volumes of re-dispatch become important and the need to start up large inflexible units for re-dispatch purposes increases, the inefficiency of dealing with re-dispatch in real time becomes more and more apparent.

In several US markets, such as California and Texas, one of the main reasons for transition to the nodal market design was the excessive cost of real-time system re-dispatch.

4.6.2 Market Power

The problem of market power in electricity markets can be exacerbated by the presence of transmission constraints. Transmission constraints create fragmented geographical markets with less competition than on a broader market level. For instance, when transmission constraints are binding in the direction of a large load centre, this centre becomes a separate market, or a load pocket. Generators outside the load pocket cannot exert competitive pressure on the generators inside the pocket, making market concentration in the load pocket higher than in the market as a whole.[14]

One of the common critiques of the nodal system is its perceived higher vulnerability to locational market power than a system based on re-dispatch. This perception is incorrect. The ability of generators to exercise locational market power depends on the topology of the transmission network, and not on the choice of congestion management mechanism. The congestion management system only determines how market power is exercised and the mechanism by which it impacts customers.

In the locational price system, market power is exercised through the submission of supply bids above variable costs with the intention of increasing locational prices. In re-dispatch systems, market power is exercised through the inflation of re-dispatch payments. This can be done by first submitting day-ahead bids that ensure that a generator participates in the re-dispatch market, and then submitting re-dispatch bids in excess of the re-dispatch equilibrium price. Since the re-dispatch costs are socialised and averaged over all system users, it is more difficult to observe the impact of the exercise of market power in the re-dispatch system.

The exercise of locational market power in re-dispatch systems involves day-ahead and re-dispatch bidding that is similar to the strategies followed by competitive generators exercising arbitrage opportunities in the presence of constraints, as described in Section 4.4, above. In both cases, generators do not bid according to their production costs. However, as opposed to competitive generators, bids by generators exercising locational market power exceed the re-dispatch competitive equilibrium price. Yet the similarity of the competitive and abusive bidding strategies makes it more difficult to distinguish the exercise of market power from competitive behaviour in the re-dispatch system than in the system with locational prices, where competitive generators bid at variable cost.

The re-dispatch approach cannot create more competition among generators by assuming no constraints in the day-ahead market. Instead, it tends to obscure the market power that is exercised in the subsequent re-dispatch market. In contrast, the locational price-based market does not create market power, but makes it more transparent.[15]

There is one important case where the re-dispatch and locational price systems may impact differently on the opportunity for generators to exercise market power. This is market power in the export-constrained area. In the re-dispatch system, a generator with such market power can set export-side re-dispatch prices at very low or negative values and achieve profits equivalent to the difference of this price from the day-ahead price. In the locational price system, such a generator would also be able to set a low or negative locational price, but there is little or no opportunity to convert this low energy price into profits.

4.6.3   Market Liquidity

It is sometimes argued that a system with locational prices may reduce the liquidity of wholesale power markets. The argument suggests that whereas in a re-dispatch system a market participant (for example, a retail supplier looking to buy power) could look for a counterparty among operators in the entire national market, in the locational price system it should either look for a counterparty within a smaller area or pay congestion price differentials if the counterparty is found in a wider area.

This argument tends to misinterpret the implications for market liquidity of both the locational price system and the re-dispatch system.

In the system of locational prices, counterparties for bilateral trades (for example, using contracts for differences) can still be found throughout the entire market. In the event of transmission congestion the transaction would incur a congestion cost equivalent to the difference between clearing prices in the sink and the source area of the trade. This congestion cost can be hedged by acquiring the FTRs between the production and the consumption nodes matching the trade. Thus, the system of locational prices does not limit the scope of bilateral trades and provides the means to hedge such trades.

The systems featuring a day-ahead market with a nationwide uniform price and subsequent re-dispatch may appear to provide more opportunities for bilateral trades throughout the entire territory. However, it is often not considered that in the event of congestion such a system can provide only a partial hedge, since no financial hedge against the volatility of the tariff component covering re-dispatch costs is generally available at market participants. Thus, in the event that re-dispatch costs are material, the re-dispatch system may impede bilateral trades because of its inability to provide a hedge for a significant wholesale cost component.

4.6.4 Implementation and Operation Complexity

A common critique of the nodal congestion management system as compared with the single-price re-dispatch approach is based on the argument that the process of calculating nodal prices in the presence of congestion is very complex and non-transparent, undermining market participants' confidence in these markets.

However, the calculation of nodal prices follows the logic of constrained optimisation algorithms, and the resulting prices meet a certain number of high-level properties, such as those presented in Section 4.2, above. Such high-level properties can be verified and audited.

Although the calculation of nodal prices is indeed more complex than the calculation of a single unconstrained clearing price, it is certainly not more complex or less transparent than the subsequent calculation of nodal re-dispatch instructions and payments that needs to be performed in the single-price system in the presence of congestion. Furthermore, as discussed in Section 4.4, above, bidding incentives induced by the re-dispatch congestion management may also affect preceding markets, such as the day-ahead market. Therefore, confining the nodal clearing system to the re-dispatch phase cannot be considered to limit the impact of its complexity to a narrow segment of the market.

4.6.5 Incentives for Network Development

It is often argued that because the system of locational prices allows the system operator to collect congestion rent, it provides incentives to the system operator to retain congestion rather than to reduce it through short-term measures or through building transmission upgrades.

However, in no existing system does the system operator appropriate the congestion rent as additional profit. The congestion rent is generally returned to market participants in one form or another. Thus the congestion rent collected by the system operator in the process of the congestion management mechanism does not have an impact on the system operator's incentives to retain congestion through short-term actions or to delay necessary transmission investments.

Regulatory incentives for the system operator to make timely investments in transmission infrastructure can be put in place regardless of the chosen method of congestion management. The regulatory incentives to invest in transmission can be based on the objective indicators of the social benefit of the investment, which does not depend on the chosen congestion mechanism.

4.6.6 Incentives for Generation Investment

Systems featuring re-dispatch congestion management and systems featuring locational prices provide different incentives for investment in generation capacity.

The re-dispatch system aims to maintain uniform price signals over the entire area. This system may give little additional incentive for generator incentives in import-constrained areas where additional capacity is most needed. On the contrary, it may induce entry in areas where there is already excess capacity and from which export is limited. New generation capacity in those areas will seldom be activated and will more often be collecting re-dispatch payments for being constrained off. This may also provide an adverse incentive for investment in inefficient plant.

Generation investment induced in re-dispatch systems further increases the need for and cost of re-dispatch.

On the other hand, systems where congestion is resolved by setting different market prices in different locations indicate the marginal value of energy in a given location given the transmission constraints. These prices provide more precise signals of the need for generation investment in different geographical areas. Locational spot-market prices can be further hedged in forward markets, and locational differences can be hedged with long-term transmission contracts, FTRs, providing market participants with the necessary instruments for mitigating investment risks.

4.6.7 Political Pressure for a Single National Price for Customers

Congestion management via locational prices implies that different customers may pay different electricity prices just because of their location. That feature has been the source of political opposition to locational prices, since price uniformity for essential goods such as electricity has traditionally been a social policy objective.

However, congestion management via locational prices may be implemented while preserving the traditional cross-subsidies among customers located in different geographical areas, through a proper market design. This happens in several US markets in which prices charged on withdrawals are uniform within each zone, and equivalent to the average of the nodal prices in the zone. In Europe the same logic applies in Italy, where withdrawals are charged at a nationwide uniform price, while generators are paid geographically differentiated prices.

As long as the electricity demand is price inelastic,[16] congestion management systems based on asymmetric prices for generation and load do not give rise to major inefficiencies, since charging loads the uniform price or the nodal price does not alter the purchased volumes. Inefficiency caused by the lack of locational signals on the demand side is more likely to appear in a longer-term horizon. Specifically, uniform demand-side prices could distort the siting decisions of large industrial customers.

4.6.8 Surplus Distribution Properties

Markets with locational energy prices and markets based on re-dispatch have very different surplus distribution properties. Whereas the locational prices reflect the marginal value of energy at each location, re-dispatch systems intend to maintain generators' revenues at the level they would be if there were no congestion. The re-dispatch system therefore creates cross-subsidisation among market areas, maintaining the price above the marginal value of energy in the area in export-constrained areas and the price below the marginal value of energy in import-constrained areas.

Furthermore, if congestion is predictable, arbitrage between the energy and the re-dispatch market may lead to a uniform market-clearing price higher than that without congestion, as we discussed in Section 4.4.2. In this case the re-dispatch system results in greater total supply costs for the consumers, compared with the nodal approach.

Although the transition from a re-dispatch market to a market with locational price differentiation can be expected to improve market efficiency overall, there may be both winners and losers as a result of this transition. Parties that may be disadvantaged by the locational system are generators located in export-constrained areas that benefit from the re-dispatch system, which allows them to maintain a price above the marginal value of energy in their location.

For these parties, transition to the locational price system can be made more acceptable by allocating them the FTRs between their export-constrained area and the remaining areas.

4.7 NETWORK DEVELOPMENT

Transmission and generation capacity are linked by complex relations of complementarity and substitutability. A network upgrade may be necessary in order to transfer a new generator's production from the injection node to the load centres. Alternatively, a transmission upgrade might make it possible to meet a demand increase at a certain location with production from existing generators at other locations, thus avoiding the construction of additional capacity near the load centre. Thus, total supply cost minimisation requires transmission and generation investment decisions to be coordinated.

In most electricity markets, the system operator is the monopoly supplier of transmission services. The system operator is responsible for planning the network's upgrades in order to increase the transmission capacity. Where generation is liberalised, private investors decide when and where new generation plants will be built. In this setting the network development decisions are not made by those who bear their economic consequences, that is, the power generators and consumers.

Coordination between transmission and generation investment decisions is different depending on the congestion management system implemented in the market. Where congestion management is carried out via re-dispatch, the economic consequences of transmission investments are borne mostly by electricity consumers. They pay transmission charges,[17] ensuring full recovery of the network costs irrespective of the actual use of the transmission capacity. This means that if the system operator's decisions lead to overinvestment, consumers pay for unnecessary transmission capacity. In the event of underinvestment, inefficiently high re-dispatch payments are made to the generators and passed on to the consumers via the transmission tariffs. As we discussed in Section 4.4, in a re-dispatch system generators do not bear any cost in the event of congestion, since the re-dispatch payments are such that constrained-off generators obtain at least the same profits as if there were no congestion.[18]

Since in this setting a generator's profit does not depend on its location,[19] the sites where new capacity is built might be inefficiently selected. In practice there are ways for the system operator to influence generators' localisation decisions, for example by denying or delaying interconnection at the nodes where additional injections would create congestion. This may reduce the scope for the inefficient localisation of generation plants. However, the transmission and generation investment decisions are coordinated by an administrative process, rather than via the pricing system.

In the event that congestion management is carried out via locational prices, both generators and consumers are affected by the system operator's network investment decisions, via the impact of network upgrades on electricity prices at the different locations, as well as via the transmission tariffs. In the event of underinvestment, electricity prices in some areas will be higher than optimal and some efficient generation capacity in some areas will not be activated (without being compensated). In the event of overinvestment consumers pay larger than necessary transmission tariffs.

In this setting the generation localisation decisions internalise the expected network upgrades via their impact on electricity prices at the different locations. For example, if the system operator's network development plan does not include upgrades relieving a certain constraint, the expectation of high prices at the import side of the bottleneck will attract new generation investment in that area. However, even in this context, coordination between transmission and generation investments is not fully market based. The system operator's views on the (efficient) future geographical distribution of the generation fleet – reflected in its network development decisions – may be different from the market's views. When this happens, there is no mechanism allowing one party to take control of the network investment decisions and assume the corresponding risk.

In an alternative approach to network development, market investors take the risk of expanding transmission capacity. They pay for network upgrades and become owners of the transmission rights that result from the upgrades. This model is often referred to as a 'merchant model', and we discuss it in the framework of a locational price congestion management system.[20] A simple example can be used to illustrate how coordination between generation and transmission decisions takes place in a merchant context. Figure 4.23 illustrates our example.

The prices shown in the figure are for a base-load future contract with delivery at each node. The price at node 1 is lower than at node 3. As we have seen in Section 4.3.2, upgrading the capacity of the line 1–3 by 100 MW allows the system operator to issue an additional 150 MW of financial transmission rights between node 1 and node 3.[21] In the merchant approach any investor would be allowed to build the network upgrade.[22]

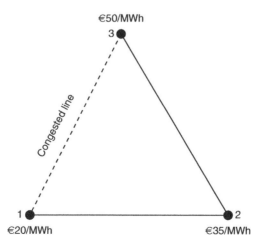

Figure 4.23   Triangular network

In exchange for each MW of the upgrade, the investor is granted 1.5 MW incremental financial transmission rights from node 1 to node 3 at no charge over the economic life of the upgrade.[23] The entire network capacity continues to be allocated on a short-term basis through the spot energy market, as shown in Section 4.3.

The investors' revenues come from exercise of the financial transmission rights. For each MW the transmission rights holder collects the price difference between nodes 3 and 1. The network capacity between nodes 1 and 3 is increased to the expected electricity prices, such that the value of a 1 MW transmission right between nodes 1 and 3 equals the incremental cost of expanding the capacity of line 1–3 by 2/3 MW.

In this context, coordination between generation and transmission investments takes place according to the standard market mechanism. Rational investors will compare the profitability of building additional generation capacity at node 1 and of increasing transmission capacity between nodes 1 and 3. Either strategy can be hedged by taking appropriate forward electricity positions. The investment in generation capacity at node 3 can be hedged by selling power at node 3. The investment in transmission capacity can be hedged by purchasing electricity at node 1 and selling the same quantity at node 3.

The lumpy nature of network investments is a potential source of inefficiency of the merchant approach. In our setting, for example, assume that the only technically feasible upgrade of line 1–3 is such that congestion is completely eliminated. After the upgrade, the electricity prices at nodes 1 and 3 will therefore be identical. Assume also that implementing the upgrade is efficient, that is, that the net surplus generated by the additional electricity transaction enabled by the upgrade is greater than its cost. Despite its positive net value, no investor would want to invest in the network upgrade, because it would result in an excess supply of transmission capacity, which would bring the value of the transmission rights from node 1 to node 3 down to zero. This is the result of the combined effect of the upgrade's lumpiness and of the impossibility of limiting the physical power flows over the network. Intuitively, once the lumpy upgrade is in place the owner cannot limit its use in order to prevent convergence of the nodes 1 and 3 prices.[24]

Furthermore, the merchant approach might not be viable if hedging the investments in network upgrades were made impossible by the lack of liquidity of the forward electricity markets.

In practice, the merchant approach could be undermined also by the difficulty of assessing the value of transmission rights granted to those investing in network upgrades. Without a detailed knowledge of the network operations, potential network investors might find it difficult to assess the impact of further transmission and generation investments on the value of their transmission rights.

In Europe a planning approach to transmission network development is implemented in all countries. Merchant cross-border investments are only allowed, as an exception to the default regime, for DC interconnectors. Exemption is granted on a case-by-case basis and only if certain conditions are met.[25] Coordination issues between merchant projects and network developments planned by system operators have not surfaced so far, presumably because of the limited scope of the merchant initiatives.

NOTES

1. To a limited extent the system operator can change the network topology, and thus the flows over network elements, by opening or closing circuits or by changing the setting of devices called 'phase shifters'. We abstract here from that possibility.
2. On how losses and non-linearity are accounted for in power flow modelling see, for example, Hogan, W.W., Pope, S.L. and Harvey, S.M. 1997. 'Transmission Capacity Reservations and Transmission Congestion Contracts', Center for Business and Government, Harvard University, 8 March.
3. For the time being we have assumed that there are no flow limits on the other lines.
4. Proponents of this approach suggest that those rights should be defined only for potentially congested network elements (Chao, H. and Peck, S., 1996. 'A Marker Mechanism For Electric Power Transition', *Journal of Regulatory Economics*, **10**, 25–59).
5. We shall discuss in greater detail the determination of the feasible set of transmission rights in Section 4.3.2, in the context of the point-to-point definition of the transmission rights.
6. Kristiansen, T., 2004. 'Markets for Financial Transmission Rights', *Energy Studies Review*, **13**(1), 25–74.
7. The same market-clearing algorithm is implemented in the real-time market.
8. CWE Orientation Study, 2008. *A Progress Report for the MoU Signatories on the Design of the Flow Based Market Coupling in the Central West European Region*, CWE MC Project, February.
9. Central Allocation Office and Consentec, 2009. *Concept of the Technical Parameters Calculation for the Flow Based Capacity Allocation in the CEE Region*.
10. Ibid.
11. Ibid.
12. The logic underlying the FRM is similar to that of the transmission reliability margins (TRMs) featuring in the traditional NTC implementation of the interface approach.
13. See, for example, Rosenberg, A.E., 2000. 'Congestion Pricing or Monopoly Pricing?', *The Electricity Journal*, **13**(3), April, 33–41.
14. In fact, constraints do not even have to be binding for locational market power to become an issue. In the event of a price increase by a generator located inside a load pocket, the mere possibility of binding transmission constraints may itself trigger the exercise of locational market power (see, for instance, Borenstein, S., Bushnell, J. and Stoft, S., 2000. 'The Competitive Effects of Transmission Capacity in a Deregulated Electricity Industry', *RAND Journal of Economics*, **3**(2), 294–325).
15. For more discussion about market power in re-dispatch and nodal systems see Harvey, S.M. and Hogan, W.W., 2000. 'Nodal and Zonal Congestion Management and the Exercise of Market Power', LECG LLC working paper.

16. At least around the equilibrium consumption level. Loads capable of responding quickly to price signals can be treated as generators in the ancillary service markets.
17. Transmission charges are typically levied on the retailers and on generators. They are then passed on to the electricity consumers.
18. One could argue that if generators implement the rational bidding strategy discussed in Section 4.4 their profits would depend on their location, which would restore the correct incentives to efficiently selecting plant locations. We do not pursue this line of reasoning here, since we want to emphasise the differences between the re-dispatch system and the nodal system that are likely to result in practice as a consequence of the various imperfections and frictions that we assumed thus far.
19. Recall that we are neglecting here transmission losses.
20. The same logic would apply to flow-gate transmission rights. A merchant approach is not consistent with a congestion management system based on re-dispatch. With re-dispatch the market perceives an unlimited supply of transmission rights, even in the event of congestion. As a consequence, there is no market for the right to use the transmission system.
21. We abstract here from the issues involved in assessing the amount of additional transmission rights that a network upgrade allows to be issued.
22. A detailed characterisation of the institutional setting is not necessary here. For example, private investors could be allowed to ask the system operator to implement the upgrade on their behalf. They would pay for the upgrade and receive the corresponding endowment of transmission rights.
23. As we discussed in Section 4.3, a network upgrade will generally allow the system operator to create alternative sets of new transmission rights. In this case the sponsors would select their preferred set of transmission rights.
24. The upgrade's sponsors could withdraw some of the financial transmission rights that they are granted by the system operator from the market. However, that would not mitigate the impact of the excess supply of physical transmission caused by the upgrade on spot energy prices.
25. The criteria for exemption from third party access for new interconnectors are laid out in Article 17 of European Commission Regulation No. 714/2009.

5. Competition policy in the electricity industry

Guido Cervigni and Dmitri Perekhodtsev

5.1 INTRODUCTION

Market power is a company's ability to profitably raise and maintain prices above the level that would prevail under competition. Market power is a primary concern in wholesale electricity markets for two broad reasons. The first is that electricity is a primary commodity purchased by every household and business, and its price is extremely important for the economy. The second is that the unique technical and economic characteristics of electricity make wholesale electricity markets particularly vulnerable to the exercise of market power. These characteristics are little or no demand responses to price changes, the fact that electricity is not storable, and tight transmission capacity constraints that reduce the scope for competition among generators connected in different locations. As a result, even small generators may have the interest and ability to induce dramatic price increases under certain conditions. The problem is exacerbated by the highly concentrated industries that have resulted in most countries from the liberalisation of electricity generation.

For these reasons, in some countries market power in the wholesale electricity market is addressed by regulatory statutes, the primary goal of which is to avoid excessive prices. In the US, for example, the Federal Power Act 1935 places a statutory obligation on the federal energy regulator to ensure that wholesale electricity prices are 'just and reasonable'. In Europe, charging excessive prices is an abuse of market power in breach of competition law,[1] although historically the prohibition on charging excessive prices has proved hard to enforce. Recent policy developments confirm the difficulty for the European authorities in addressing market-power issues in wholesale electricity markets in the current legal framework. For example, in June 2012 the Austrian government submitted a new cartel law to parliament, which turns around the burden of proof in cases of abusive high prices in the energy sector only. Under the new law, electricity and gas companies will have to justify their prices if they are higher than in comparable markets.[2] In the UK, certain pricing policies by generators in case of congestion – which do not violate the competition rules – have been banned through an *ad hoc* licence condition.

In the electricity generation market, more frequently than in others, the regulatory authorities implement structural or quasi-structural measures in order to mitigate market power, or enforce specific *ex ante* restrictions on companies' behaviour, such as price caps, bidding restrictions or mandated cost-based offers. These measures complement the deterrence of harmful conduct through the prospect of *ex post* investigations and fines for breach of competition laws.

The standard competition policy toolbox needs to be tailored to the specific technical features of electricity supply. In particular, the non-storability of electricity, demand price inflexibility and network constraints make application of the standard methodology to determine the market boundaries – the hypothetical monopolist test – less than straightforward. Furthermore, as the competitive interaction among generators takes place in auction-based markets, the usual concentration measures based on market shares may be poor indicators of market power in electricity generation.

In Sections 5.2 and 5.3 we discuss the methodology for assessing competition in wholesale electricity markets based on market structure (5.2) and the generator's market behaviour (5.3). In Section 5.4 we analyse the main market-power mitigation measures implemented in wholesale electricity markets. Finally, in Section 5.3 we present a selection of competition policy cases.

5.2   MARKET POWER ASSESSMENT BASED ON MARKET STRUCTURE

The assessment of competitive conditions and market power is a key element of antitrust and merger control cases. Certain business practices are regarded as abusive if implemented by a firm enjoying considerable market power, or in a dominant position, while being allowed for non-dominant firms. A merger between firms operating in the same market may not receive approval from the competition authorities, or be approved subject to conditions, if it is likely to restrict competition.

A firm is said to have market power if its customers and competitors exercise few constraints on its actions, so that the firm has the ability and the incentive to raise and maintain the price above the level that would prevail under competition. Figure 5.1 illustrates the concept of market power. Both panels of the figure show the firm's competitive supply: in

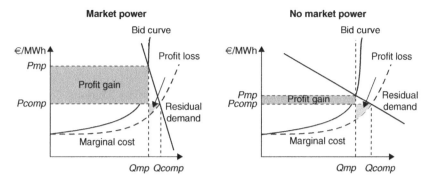

Figure 5.1   Ability and incentives to exercise market power

a competitive environment the firm will increase sales until the cost of producing an additional unit equals the price. In other words, the firm's competitive supply coincides with the firm's marginal cost function.

The figure also shows the residual demand faced by the firm. For each price level, the residual demand represents the market demand net of the volumes supplied by the firm's competitors. The residual demand summ- arises the pressure exercised on the firm by the supply of competing gen- erators and by the reaction of customer demand to price changes.

In the figure we assume for simplicity's sake that the entire supply of the firm is submitted in the spot market in the form of a bid curve. The firm's bid curve and residual demand cross at the market-clearing price. Setting the price above the competitive level has two contrasting effects on the firm's profits: on the one hand a profit loss because of the decrease in the volumes sold, and on the other a profit gain because of the increase in the margin obtained on sales at the higher price.

The left panel of the figure illustrates a situation where the firm has market power. A price increase above the competitive level brings about a net increase in the firm's profit. The additional margin obtained on the volumes sold at the higher price outweighs the negative impact of the reduction in sales.[3] The right panel illustrates a situation without market power. The same bid curve induces a small price increase, and the profit gain on the volumes sold is insufficient to outweigh the profit loss caused by the reduced sales.

Different slopes of the residual demand, which determine whether or not the firm has market power, represent different strengths of the com- petitive constraints of competitors and the price elasticity of demand.

Below we discuss the nature of competitive constraints in wholesale electricity markets in greater detail. We then present two approaches to assessing the degree of competition in wholesale electricity markets: the traditional approach applied to a wide variety of markets in competition policy and an alternative approach that directly assesses competitive constraints.

5.2.1 Competitive Constraints in Wholesale Electricity Markets

The market power of an electricity generator can be mitigated by the reaction of the consumers and of competing generators to a price increase. We discuss these in turn in the following subsections.

Electricity demand elasticity

Electricity demand is typically considered very inelastic. This is partly because electricity is the only available energy source for most uses, and can only be substituted by other fuels such as natural gas or fuel oil in a limited number of cases. However, such substitution involves replacing appliances and should be regarded as long term rather than a possible demand response to a price change in the short run. Therefore it is commonly considered that short-run demand-side substitution provides few constraints on the market behaviour of firms operating in the wholesale electricity markets.

As discussed in Chapter 2, much of the electricity produced and consumed is traded on forward markets, with various time horizons. Transactions are executed on exchanges, trading platforms and bilaterally. The question often arises as to whether electricity volumes sold in different trading venues and with different forward horizons are substitutes. When wholesale markets are liquid, one would expect market participants to be able to arbitrage freely across the different products and trading venues. A sustained increase in prices on the spot market would induce a shift of demand away from spot to forward products, which would also increase forward prices. A price increase for products sold on a power exchange would lead to a shift in demand towards bilateral trades, which would result in an equal price increase. On this basis it is commonly considered that all transactions involving the production and consumption of electricity at a certain time are good substitutes and therefore belong to the same market.[4]

However, real-time and re-dispatch transactions are, at least in the European context, typically not considered to be part of the wholesale electricity market. This is the consequence of several features of the design of the European electricity markets, including the following:

- small generators and all but very large consumers are generally not allowed to participate in the re-dispatch and real-time markets; and

- the imbalance pricing system in most countries makes voluntary imbalances (a way to trade on the real-time market) unprofitable, irrespective of the actual cost (or cost reduction) for the system operator.

Competition among generators

If a firm operating in the wholesale electricity market increases its offer prices, competing generators will get to produce more. In wholesale electricity markets this displacement takes place immediately, thanks to power exchanges that clear bids and offers according to their price merit order.

The extent of competitive constraints may vary hour by hour depending on the demand level, the corresponding set of competing generators and fuel prices. In Figure 5.2 we illustrate how demand level impacts on the competitive constraint faced by a generator. We consider the competitive situation in off-peak and peak hours. On the supply side, we assume that only two technologies are available, with materially different variable cost (*VC* in the figure): base-load technology with low variable cost, and peak technology with a high variable cost.

We can now assess the market power of a firm controlling a large portion of the base-load capacity but none of the peak capacity. In this case the competitive pressure faced by the firm is very different in peak and off-peak hours. In off-peak hours the firm enjoys significant market power. It finds it profitable to set a price just below *VC Peak*, as shown in the figure, well above the competitive level *VC Base-load*. However, in peak hours, when the market price is set by the variable cost of peak

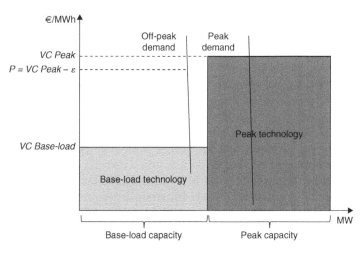

Figure 5.2   Competition for different demand levels

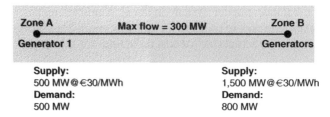

Figure 5.3   Market power and transmission constraints

technology, the competitors controlling peak capacity make it unprofitable for the firm to set a price above the competitive level *VC Peak*, because doing so would require a large reduction in the volumes sold.

Note incidentally that, since electricity is non storable on a massive scale, it is impossible for electricity produced at one time to replace electricity produced at another time, or in other words for the generation capacity in service at one time to compete against the generation capacity in service at another time. Furthermore, the cross-elasticity of demand at different times appears to be very limited: very few consumers would postpone electricity consumption in the event of a four-hour price increase, for example. As a consequence, the decision to consider large sets of hours as part of the same market is based not on supply- or demand-side substitutability across hours, but on the assessment that very similar competitive conditions hold in all hours.[5]

The displacement of a generator increasing the offer price by generators connected to the transmission network at different locations can be limited by the presence of transmission constraints. Consider the example of two areas, illustrated in Figure 5.3. In this figure Generator 1 with a capacity of 500 MW and demand of 500 MW is located in zone A and generators with a combined capacity of 1,500 MW and demand of 800 MW are located in zone B. We also assume that the cost of all the generators is identical.

If transmission capacity connecting the two zones were unlimited, the generators located in zone B could exert strong competitive pressure on Generator 1. A slight increase in the offer price of Generator 1 would place its capacity out of merit, and generators from zone B would supply the demand in both zones A and B.

However, a constraint of 300 MW on transmission capacity between the zones could dramatically reduce the competitive pressure from generators in zone B on Generator 1. The available transmission capacity would only allow the generators from zone B to meet 300 MW of the demand in zone A, and the remaining 200 MW of demand would have to be met by Generator 1, allowing it to set a higher offer price in its zone.[6]

5.2.2 The Standard Approach to Market-power Assessment

In competition policy, market-power assessment is generally carried out in two steps.[7] First, the boundaries of the relevant market or markets are identified (market definition). Then the degree of competition is assessed through concentration indices based on the market shares of the firms belonging to the relevant market identified in the first step (competitive assessment). Below we discuss these steps of the standard approach and its weaknesses when applied to wholesale electricity markets.

Market definition analysis
According to the European Commission:

> Market definition is a tool to identify and define the boundaries of competition between firms. . . . The objective of defining a market in both its product and geographic dimension is to identify those actual competitors of the undertakings involved that are capable of constraining those undertakings' behaviour and of preventing them from behaving independently of effective competitive pressure.[8]

The standard conceptual framework for market definition is the hypothetical monopolist test, or SSNIP test (small but significant non-transitory increase of price). The test starts with identification of the smallest conceivable candidate market, that is, the smallest set of products and locations that are believed to be highly substitutable and therefore to belong to the same market.

The second step of the test consists of assessing the competitive constraints that the products and firms external to the candidate market could exercise on the products and firms included in the candidate market. This is done by assessing whether a hypothetical monopolist in the candidate market would find it profitable to implement a small (typically 5–10 per cent) price increase for a relatively long period (such as one or two years). Such a price increase is less profitable the more it results in:

- customers replacing the products supplied by the hypothetical monopolist with products not included in the candidate market – demand-side substitution – or
- other firms switching their production facilities to produce the products included in the candidate market – supply-side substitution.

For example, in the retail car market the initial candidate market could include all makes of city cars sold in a certain city. The profitability of the price increase by a hypothetical monopolist in this candidate market would depend on:

- The price elasticity of the demand for city cars. Following the price increase the demand for city cars would reduce as the customers substitute it with other products. For example, an increase in the price of city cars might lead consumers to buy fewer city cars and more medium-sized cars. Alternatively, consumers could travel to neighbouring cities not included in the candidate market to buy city cars. These are instances of demand-side substitution.
- The behaviour of retailers operating at other locations or retailers of different types of cars operating in the candidate market area. The former could find it profitable to respond to the price increase by opening branches selling small cars in the candidate market. The latter could expand their offering of city cars in the candidate market at little cost. These are two examples of supply-side substitution.

If demand- and/or supply-side substitutions are major enough to make the price increase unprofitable for the hypothetical monopolist, this indicates that the products included in the candidate market are subject to competition from some of the products excluded from the candidate market. In this case the profitability of a price increase by a hypothetical monopolist is tested for a larger candidate market, including the product and/or location that are the best substitutes for those making up the first candidate market. The candidate market is enlarged and the test is repeated until the price increase turns out to be profitable. The set of products and geographical locations found in this way satisfy the requisites that define a market: they exercise significant competitive pressure one on the other, while facing limited competitive pressure from those external to the set. In the car example, the candidate market might have to be expanded to include all the small and medium-sized cars sold in the entire country.

The application of SSNIP logic to assess the boundaries of wholesale electricity markets along the product, geographical and temporal dimensions reflects the nature of the competitive constraints in the market discussed above. When markets are sufficiently liquid, a high degree of substitution between spot and forward electricity transactions executed bilaterally and on exchanges is to be expected.[9] For this reason such transactions are often considered as belonging to the same market, and the competitive assessment is then carried out as if all electricity produced and consumed were exchanged on a spot basis, under the assumption that arbitrage between the spot and the forward markets takes place.[10]

However, in the European context, real-time and re-dispatch transactions are often considered a separate product market.[11]

Second, in electricity markets competitive conditions may vary greatly between different hours. This is because different demand levels typically correspond to different sets of potentially competing generators. In addition, the generation merit order may change as a result of changes in prices of fuels and CO_2 permits. For this reason several distinct relevant markets may be identified at different levels of demand. The US Federal Energy Regulatory Commission (FERC) implements various tests in order to assess the market power of generators applying to be exempted from price regulation.[12] One of these is the delivered price test, which identifies 'potential suppliers based on market prices, input costs, and transmission availability'.[13] FERC considers 10 separate periods of the year, potentially featuring different structural conditions.[14] For each period, FERC determines a representative market price and identifies the relevant market as the set of generators with variable costs lower than 105 per cent of the reference variable cost. Competitive assessment is then carried out for that market.[15] The Brattle Group has investigated the wholesale electricity market definition in the Netherlands in the context of merger control.[16] On the basis of the observed price differentials they recommend considering three separate markets: off-peak, peak and super-peak hours.[17]

Finally, transmission constraints may determine boundaries to the relevant geographical markets for wholesale electricity. For instance, when transmission constraints are binding in the direction of a large load centre, this centre may become a separate relevant market, or a load pocket, because generators located outside of the load pocket may not exert any competitive pressure on the generators within the pocket. The geographical dimension of the relevant market may vary in different sets of hours because the level of demand has a major impact on network congestion and the resulting geographical boundaries of the wholesale electricity market.[18]

Competitive assessment
Once the relevant market has been identified, the standard approach assesses market power indirectly, through indices based on the market shares of firms operating in the relevant market. Measuring market power through concentration indices has an intuitive appeal. Other things being equal, one would expect a monopolist with a 100 per cent share to have the highest possible market power and a very small firm to be unable to exercise market power. Competition authorities rely on market-share thresholds to judge market power. A firm with market share in excess of 70 per cent is presumed to be dominant. A share between 50 and 70 per cent

raises a weaker presumption of dominance. A market share below 40 per cent is commonly regarded as unable to support a finding of dominance.[19]

A widely used index of market concentration based on market shares is the Hirschman–Herfindahl Index (HHI). The HHI is calculated as the sum of squares of the market shares s of the suppliers in the market.[20] In European case law, HHI levels above 2,000 are generally considered to indicate a concentrated market, whereas HHI levels below 1,000 indicate a non-concentrated market. In the US electricity market, FERC considers that a generator does not raise market-power concerns provided that the HHI is below 2,500 and the generator's market share is less than 20 per cent (in production capacity), or the generator's market share is above 20 per cent and the HHI less than 1,000.

The wide use of market share and especially the HHI to assess competition in the relevant market is supported by the theoretical models most commonly used to describe the strategic interaction between imperfectly competing firms in the market. For example the Cournot model,[21] the workhorse of competition analysis, establishes the following relationship between the HHI and the price mark-up over the marginal cost resulting from the imperfect competition in this market:

$$\frac{p - c}{p} = \frac{HHI}{\varepsilon},$$

where ε is the elasticity of demand in the relevant market. As the equation shows, the price-cost mark-up – the measure of market power – is directly related to the HHI if the interaction between the firms in the market is well described by the Cournot model.

Issues with application of the standard approach to wholesale electricity markets

The standard approach to market-power assessment is widely used and applied to a variety of markets because of its simplicity. Calculation of market share within an identified relevant market is straightforward and requires only very basic data. Quantitative implementation of the SSNIP test may still be quite a data-intensive and complex exercise, but in practice it is often performed approximately using high-level qualitative information.

Nevertheless, application of the SSNIP test to wholesale electricity markets presents specific issues related to the unique technical features of electricity. First, the assumption underlying the reference to a 5–10 per cent price increase in the SSNIP test is that if the 5–10 per cent price increase turns out to be unprofitable for the hypothetical monopolist, a larger price increase will be all the more unprofitable as it will cause more intense demand- and supply-side substitution. This may not happen on the wholesale electricity market. The relationship between the price increase and the hypothetical monopolist's profit may not be monotonic, because once the competitors' capacity is fully utilised, both demand and the competitors' supply are completely price inelastic in the short term.

Second, because of transmission constraints the boundaries between the different geographical markets may not be symmetrical. For example, consider a merger between some generators in area A in the basic two-area network shown in Figure 5.3. In order to identify the relevant market the SSNIP test would be run, starting with area A, to verify if the generators in B would provide an effective competitive constraint for the generators in A. We assume that the SSNIP test leads to the conclusion that the generators in A and in B belong to the same market. However, the SSNIP test that determines whether generators in B are an effective competitive constraint for those in A does not necessarily imply that generators in A are an effective competitive constraint for those in B. Depending on the demand and supply conditions in A and in B, the same transmission capacity might be enough to discipline a hypothetical monopolist in A, but not in B. As a result, the market definition will be different depending on whether the first step of the SSNIP test uses area A or area B as a candidate market.

Third, as we discussed extensively in Chapter 4, in real, highly meshed transmission networks the relationships of complementarity and substitutability between production at different nodes may be complex and unstable. As a consequence, identifying the sets of generators to include in a candidate market within the context of the SSNIP test may be a daunting exercise. The problem is exacerbated by the possibility that the outcome of the SSNIP test could be path dependent. For example, it could be that, (i) by adding a certain portion of network x to a starting candidate market, the SSNIP test concludes that a market has been found and the investigation stops; but (ii) the same conclusion would have been reached had a different portion y of the network been added to the initial candidate market instead of portion x. In other words, the boundaries of the market that is ultimately identified through the SSNIP test depend on the order in which additional areas of the network are considered.

The specific features of the electricity market also impact on the methodology used to assess the level of competition. The use of the HHI to assess market power is justified by the Cournot model of oligopoly. This model predicts that the market price will be proportional to the HHI, and inversely proportional to the price elasticity of demand. However, the model fails to provide meaningful results when applied to electricity markets with nearly zero demand elasticity.[22]

Furthermore, trading in the spot wholesale markets takes place through auctions, and a substantial body of economic analysis focuses on the generators' optimal bidding strategy in that framework.[23] These analyses suggest that the Cournot model is not suitable for describing the interaction between generators in the auctions and that type of market power. As a consequence, the relationship between market structure and market power, on which the assessment of market power through concentration indices is based, appears not to hold for wholesale electricity.

5.2.3   Market-power Assessment through Structural Models

A more theoretically rigorous approach to assessing market power in wholesale electricity markets is through an equilibrium market model. In this approach, a model is identified that is believed to accurately represent strategic interaction between the generators on the market. The model parameters are estimated based on the observed market variables. The departure of the observed market outcomes from those that would prevail under perfectly competitive conditions is then assessed on this basis. The model is also used to predict the prices that would result from a concentration in the industry.[24]

Equilibrium models are appealing for their theoretical basis, as they consider the offer strategy of each generator to be the maximum-profit response to other generators' offers. However, most such models find multiple results in terms of possible market equilibrium. This reduces the model's ability to predict the market outcome under market conditions different from those on which it has been estimated, as would happen if the model were used to assess the effect of a merger. In addition, market equilibrium models are typically based on a highly simplified representation of the generators' cost function, and neglect the links between offer strategies that come into play at different times. These drawbacks, along with the intrinsic complexity of the approach, limit the use of equilibrium models in regulatory and antitrust proceedings.

5.2.4   Direct Methods to Assess Market Power and the Pivotality Approach

Alternative approaches to assessing market power have been developed to overcome some of the standard methodology issues. For example, direct approaches may involve estimating the residual demand (illustrated in Figure 5.1, above) faced by a given firm each hour. In areas where bidding data are publicly available and the bids can be associated with specific generators, residual demand elasticity can be explicitly estimated from the bids of competitive generators. The residual demand is given by the market demand after subtracting the supply bids of all the other participants and the net import schedules. Analyses of the residual demand elasticity were performed in order to rule on market concentration in the electricity markets of California, New Zealand and Italy.[25] However, the data required to directly estimate the residual demand are not always available. In such cases, a simplified analysis of residual demand can be performed, known as a 'pivotality analysis'.

Measuring generators' pivotality

Pivotality measures the ability of a particular generator, or group of generators, to set the market price when all other suppliers bid competitively. For example, in a market in which the total generating and import capacity is 10,000 MW and the demand in a given hour is 8,000 MW, a generator controlling more than 2,000 MW capacity is indispensable, or pivotal, since demand cannot be met without using at least part of that generator's capacity. Given the extremely low elasticity of the demand for electricity, this operator is then able to command any price for its output, up to the value of lost load, or VoLL. Figure 5.4 illustrates the notion of pivotality.

In the figure we assume for simplicity's sake that all generators have the same variable cost and we assess the competitive position of Generator 1. Some of the capacity controlled by Generator 1 is necessary to meet load even if the capacity of all the other generators and the import capacity are fully used. Generator 1, then, is pivotal. Given that demand is inflexible

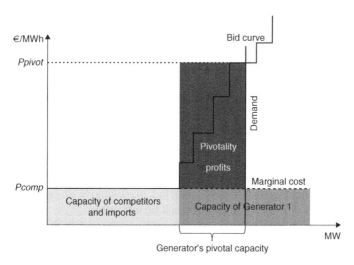

Figure 5.4   Generator's pivotality

in the short run, Generator 1 has the ability to set the market price at an arbitrarily high level.[26] The minimum capacity of a generator that needs to be scheduled to avoid an outage is its pivotal capacity.

Several indices of market power have been developed based on the concept of pivotality.[27] They assess a generator's market power according to the number of hours in which the generator is pivotal over a long period such as a year, and/or on the generator's pivotal capacity.

The assessment of market power based on pivotality indices does not require an explicit preliminary assessment of the geographical boundaries of the market. The pivotality calculation takes into account the competitive pressure exercised by generators located outside the area for which the generator's market power is being assessed, by assuming that the entire transmission capacity can be used to import electricity.

We discuss below the limitations of the assessment of market power based on pivotality and possible adjustments to the simpler pivotality analysis.

An extreme model of competitive interaction
The pivotality indices reflect a very specific scheme of competitive interaction, in which market power is exercised only when the generator is necessary in order to satisfy demand. Specifically, pivotality measurements do not take into account the fact that even non-pivotal generators may have some control over prices, since by withholding their capacity they force higher cost units to set the price. Large generators might find it profit maximising to use this strategy, even if they are not pivotal. This is the case, for example, of the generator controlling a large part of the base-load capacity in Figure 5.2, which can set the price at a level immediately below the cost of the peaking units even without being pivotal. This possibility is not captured by the basic pivotality indices.

Technical constraints
Standard pivotality measurements are generally based on the assumption that the generator can withhold up to its entire capacity at any time in order to increase the market-clearing price. In reality this may not always be feasible. For various technical or economic reasons, some types of electricity generation such as nuclear, hydropower, wind, and combined heat and power commonly cannot freely modify their output. Withholding the capacity of those units from the market, particularly at selected hours, may engender considerable costs or just be infeasible.

In the pivotality calculation, one can take into account the limited possibility of withholding must-run units by subtracting inflexible generation capacity from demand (Figure 5.5). Thus the pivotal capacity of Generator 1, after taking into account its inflexible capacity, is:

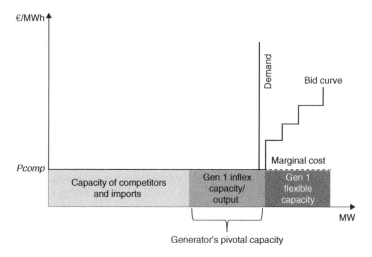

Figure 5.5   Pivotality adjustment for inflexible capacity

$$Pivotal\ capacity\ (ability)_1 = Demand - Capacity_{-1}$$
$$- All\ non\text{-}flexible\ output_1,$$

where *Capacity*$_{-1}$ is the total capacity of all the generators except 1 including total net import capacity, and *All non-flexible output*$_1$ is Generator 1's production from must-run power plants.

Incentives to increase prices
Pivotality measures the ability of generators to raise prices but does not explicitly address the incentives to do so. Pivotal behaviour may entail a dramatic loss of sold volumes; in this case the pivotal strategy only turns out to be profitable if it induces an extremely high price. However there may be explicit or implicit limits on the price increase, such as the VoLL or lower regulatory price caps. A ceiling on the market-clearing price could render a price increase unprofitable for a pivotal generator if only a small share of its capacity is indispensible to meet demand.[28]

Furthermore, the pivotality calculation does not take into account the fact that the generator's incentive to increase the spot-market price may be limited if a significant share of its capacity is already committed to production at a fixed price. For example, this may be the result of long-term sales or a commitment to serve the generator's own final customers. In other terms, incentives to exercise market power depend on the generator's net position on the wholesale market.

The pivotality calculation can be adjusted to take into account the net

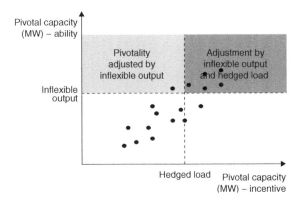

Figure 5.6   Adjustments to the pivotality analysis

position of the generator. For the purpose of assessing whether long term sales have an extreme impact on prices, the pivotal capacity of Generator 1 can be calculated as:

$$Pivotal\ capacity\ (incentive)_1 = Demand - Capacity_{-1} - Hedged\ capacity_1,$$

where *Hedged capacity*$_1$ is the share of Generator 1's capacity hedged by regulatory or contractual obligations.

Figure 5.6 shows examples of market-power assessments based on pivotality, adjusted for both flexibility and incentives. On the vertical axis we measure the generator's ability to exercise market power through the *Pivotal capacity (ability)* index. On the horizontal axis we measure the generator's incentives to exercise market power, through the *Pivotal capacity (incentive)* index. Each dot in the figure represents the pivotal capacities for the generator in one hour. The main source of differentiation between hours is the level of demand, but also the available capacity of competing generators and import capacity in different hours.

A generator that is pivotal when taking into account its total capacity, but whose pivotal capacity adjusted by must-run capacity and hedged load falls in the lower-left quadrant has neither the ability nor the incentive to have an extreme price impact in that hour. The upper quadrants represent hours where a generator has the ability to impact price using its flexible capacity. Finally, in the upper-right quadrant the generator has both the ability and the incentive to have an impact on price. The overall pivotality of the generator can be assessed by the number of hours of the year when the generator has both the ability and the incentive to be pivotal.

Pivotality benchmarks

The pivotality analysis is becoming part of the standard toolbox of competition analysis as a first screening for market-power issues in wholesale electricity markets. However, no widely agreed standard is yet available regarding which level of pivotality should be regarded as an unambiguous signal of dominance or significant market power.

In the US, FERC employs two tests for the absence of market power: one based on market share and one based on pivotality. If a generator has less than a market share of 20 per cent and is not pivotal, FERC makes a rebuttable presumption that it does not possess significant market power, a necessary condition for lifting the cost-based bidding obligation on the generator.[29] Some US markets employ pivotality-based tests as a trigger for market-power mitigation based on the bid caps.[30]

A competitive benchmark based on pivotality was proposed by the Market Surveillance Committee (MSC) on the California electricity market. It suggested using a pivotality-related Residual Supplier Index, or RSI, calculated in a given hour as:

$$RSI_1 = (Total\ capacity - Generator\ 1's\ capacity)/Total\ demand.$$

A pivotal generator thus has an RSI of less than 1; the smaller the value of RSI, the more dependent the market is on the capacity of this generator, and thus the more market power the generator has. MSC has recommended that for a competitive electricity market the RSI should not be less than 1.1 for more than 5 per cent of the hours in a year.[31]

Another example of a competitive benchmark using pivotality measurements was proposed by the Italian antitrust authority (AGCM) during its investigation of a possible abuse of dominant position by Enel in 2005 and 2006.[32] To identify whether Enel was dominant, AGCM calculated an index of Enel's pivotality in four macro zones[33] of the Italian market where it was present. AGCM found that Enel was pivotal in 100 per cent of hours in the Macro South; 44 per cent of hours in the North; 29 per cent of hours in Sardinia; and 24 per cent of hours in Macro Sicily.

In addition AGCM considered cases in which Enel is not pivotal in a given market but is still pivotal by virtue of its simultaneous presence in more than one macro zone, impacting the flows between market zones. Applying this analysis, AGCM found that Enel was pivotal in each pair of macro zones: 95 per cent of hours in the North/Macro South, 91 per cent of hours in the Macro South/Macro Sicily, and 63 per cent of hours in Macro South/Sardinia.

Based on this analysis, AGCM concluded that Enel has extensive market power in all the macro zones.

5.3   MARKET POWER ASSESSMENT BASED ON GENERATOR'S MARKET BEHAVIOUR

The structural measures discussed above provide information about firms' ability and incentives to impact the market price. However, they are not sufficient to identify whether market power has in fact been exercised.

Below we discuss various screening tests that have been developed to assess competitive behaviour in electricity markets. We also discuss the problems of estimating the variable cost of individual generators and the system marginal cost. Finally, we present a short review of studies of competitive behaviour in different countries.

5.3.1   Screening Tests

Economists and regulators have developed and implemented various screening tests in order to assess the exercise of market power in electricity markets. Generally these tests are based on comparison between the actual behaviour of firms and the actual market outcome with those that could be expected in a competitive environment where generators are expected to offer their capacity at variable cost, and market price is expected to be set at the marginal cost of energy in the system.

For example, a series of studies has analysed generators' behaviour, comparing the actual prices observed on the market with the competitive price estimates based on variable production costs. The costs of generators were assessed based on average heat rates and fuel prices, and together with generators' capacity were used to construct a short-run market marginal cost curve by stacking generators from the least expensive to the most expensive. The curve represents the supply curve that should prevail under competition. The hourly competitive price estimates were then obtained by intersecting these supply curves with hourly demand. To add realism to this modelling, import supply, operating reserve requirements and the probability of unit outages are also taken into account.[34]

Other tests involve analysing the data on generators' offers and sales. One such test, known as 'output gap analysis', checks whether at the observed market-clearing price a given firm has committed its production units efficiently. The test specifically verifies whether the units that have not been committed would have profitably produced at the observed market-clearing price. Failure to commit economic units (that is, presence of the output gap) can be considered capacity withholding aimed at increasing wholesale energy prices.

Another test using generator-specific bidding data involves comparing the level of the generator's offer curve in the market with the marginal cost of the generator's portfolio. The test verifies whether there are systematic deviations between the bidding curve submitted for the operator's generating capacity and the competitive marginal cost benchmark.

5.3.2   Estimation of the Competitive Benchmark

Tests of competitive behaviour in the electricity markets involve assessing the variable cost of energy produced by individual plants as well as the marginal cost of energy in the entire system. Electricity markets may seem to provide a large amount of data necessary for this exercise. The data on heat rates and the maximum production capacity of generating plants are often available. Information on fuel costs, which represent the bulk of variable production costs, can be obtained from price indices available publicly or from private reporting services. Thus, unlike other industries, estimating the costs of electricity generation would seem to be a straightforward and precise exercise.

However, even with these data, one may still fail to account for a large number of complexities of the electricity markets, which may lead to significant underestimation of the competitive generator offers as well as of the competitive market price. Cost-estimation approaches are usually static and do not take into account inter-temporal constraints faced by the generating units, such as ramp rates and minimum-run time constraints, or the unit commitment costs, such as start-up costs and minimum load costs.

Some units may face constraints on the total amount of energy that they can produce over a relatively long period, such as a month or a year. These are typically hydropower units with large reservoirs and thermal plants that operate under various environmental constraints. These units need to optimise the use of a limited amount of available energy over a long period, and to allocate production to the hours during that period with the highest prices. For energy-limited generators, producing in a given hour may entail forgoing profits for sales in other hours, that is, an opportunity cost. Therefore the profit-maximising (competitive) offers by energy-limited units generally depart from their variable production cost. Note that assessment by generators of opportunity costs considers the profits that would be obtained by selling their energy endowment at a different time, that is, the future evolution of the market-clearing price.[35]

All these details of the actual production constraints and costs impose significant restrictions on the combined production possibility frontier of generators. The actual marginal cost of energy may be considerably higher than the marginal cost estimated without taking such restrictions into account.

Finally, unlike the US markets, in most European electricity markets the market bid format is simple, and the clearing mechanism does not take into account the entire set of operational constraints of power plants. In these markets, accurate assessment of the costs and competitive behaviour of generators would involve using the actual operational data from portfolio optimisation operations performed by generating companies.

5.3.3   International Comparison of the Reference Price-cost Margin

Studies of market-power behaviour are often based on the relative margin between the observed market price and competitive price benchmark which is often assumed to be equivalent to the energy marginal cost estimate. In Table 5.1 we present a summary of several such behavioural metrics performed by competition authorities, energy regulators and as part of academic studies in Europe and the United States.

Table 5.1   International comparison of price-cost margin estimations

Market	Price-cost margin (%)	Period	Source
Germany	27	2003–2005	EC Energy Sector Inquiry 2007[a]
Spain	21		
Netherlands	6		
UK	11		
UK	20–25	1992–1994	Wolfram 1998[b]
Germany	21	2004 peak	Hirschhausen and Weigt 2007[c]
Italy	10–20	2004 peak	Perekhodtsev and Baselice 2008[d]
PJM	3	2008	US wholesale market monitoring
California	4	2008	reports[e]

Sources:
a   *Enquête menée par la Commission européenne en vertu de l'article 17 du règlement (CE) n° 1/2003 sur les secteurs européens du gaz et de l'électricité (rapport final)*, 10 January 2007, {SEC(2006) 1724}, COM(2006) 851 final.
b   Wolfram, C., 1999. 'Measuring Duopoly Power in the British Electricity Spot Market', *American Economic Review*, **89**(4) September, 805–26.
c   Weigt and Von Hirschhausen, 2007 (see n. 34).
d   Perekhodtsev, D. and Baselice, R., 2008. 'Measuring Competitive Behaviour in the Italian Power Exchange', paper presented at the 31st IAEE International Conference, Istanbul, Turkey, 18–20 June.
e   Department of Market Monitoring CAISO, 2008. *Market Issues and Performance*; Annual Report Monitoring Analytics, LLC, 2008, State of the Market Report for PJM.

The table suggests a wide range of positive margins and may indicate a variety of competitive conditions. However, it should be borne in mind that these studies represent very different approaches to evaluating the competitive benchmark and estimating the system marginal cost. In estimating the competitive benchmark, the studies range in accuracy from evaluation of the system marginal cost from the static supply curve estimated from the national generating portfolio data, to the studies performed by the market monitors in the US where the competitive benchmark was estimated from the detailed data on generators bids, and the format of bids provides complete information about production constraints on plants.

Regulators do not usually use these margins as a trigger for intervening and taking action against abusive market behaviour. An exception is provided by the MSC of the Californian market, which suggested using a 10 per cent threshold of the average annual margin. If the margin is below this level, the MSC considers that prices are just and reasonable and that the risk of market-power exercise is limited. It would suggest a regulatory intervention if the price-cost margin exceeded 10 per cent. During the California power crisis in 2000, when high wholesale prices induced the regulator to intervene, the average price-cost margin reached 40 per cent, that is, far above the threshold.

5.4   MARKET-POWER MITIGATION MEASURES

In some countries, liberalisation of the wholesale electricity markets has been accompanied by the introduction of market-power mitigation measures. In this section we discuss the main types of market-power mitigation mechanisms.

5.4.1   Divestiture, Contracts and Virtual Power Plants

Generation capacity divestiture
Divestitures were implemented in some European countries when electricity generation was liberalised – as in Italy and the UK in the 1990s – in order to mitigate the market power of the former monopolist.[36] Generation capacity divestitures have also been accepted as a remedy by competition authorities in several cases of mergers and abuse of dominance.[37]

Figure 5.7 illustrates the impact of the divestment of generation capacity by a dominant firm facing competition from a price-taking fringe. Recall that the residual demand function represents the demand for electricity net of the fringe's production, at each price level. Divestiture of

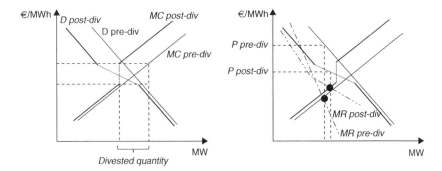

Figure 5.7   Impact of generation capacity divestment

the dominant firm's capacity has three effects. First, it increases the dominant firm's variable costs. Part of the marginal cost curve of the dominant
firm translates upwards, as the segment corresponding to the divested
capacity is removed. Second, the residual demand for the dominant firm
reduces, as the divested generation capacity is offered in the market by
the new owners. This effect is shown in the figure as a translation to the
left of the residual demand. Third, the divestiture changes the slope of the
residual demand, since the divested capacity is offered competitively – at
prices equivalent to the variable costs – by the fringe.

The right panel of the figure shows the pre- and post-divestiture market-
clearing price. The dominant firm selects the profit-maximising volume
and quantity by behaving as a monopolist on the residual demand curve,
that is, at a level such that the revenue from selling an additional unit and
the incremental cost of supplying it are equal. As a consequence of the
divestiture, the profit-maximising price for the dominant firm therefore
reduces.

Federico and López (2009),[38] show that the divestiture's impact on the
market-clearing price depends on both the size and the position of the
divested capacity on the dominant firm's cost curve. They show that
the largest price reduction is achieved by divesting capacity with marginal cost close to the post-divestiture market-clearing price. The most
effective divestment policy entails transferring capacity that: (i) in the
pre-divestment equilibrium would not be used, because offered by the
dominant firm at a price above the variable cost, and (ii) in the post-
divestment equilibrium would be activated, once offered competitively by
the fringe.

As shown in the figure, the optimal divestment policy brings the
dominant firm to select the profit-maximising price on the portion of the

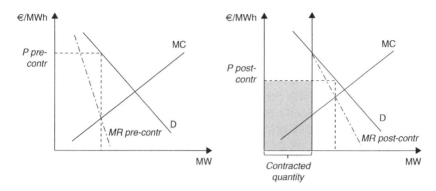

Figure 5.8   Impact of long-term contracts

residual demand curve that was flattened by the transfer of assets to the fringe.[39]

Long-term contracts

A long-term contract commits the generator to deliver a certain production volume at a predetermined price at a future date.[40] Figure 5.8 shows how a forward sale impacts on the dominant firm's profit-maximising strategy in the spot market. The spot-market equilibrium that would result if the dominant firm were not contracted is shown in the left panel. The long-term contract makes the firm's revenues from the contracted quantity independent of the spot-market price. In other words, the long-term contract simultaneously reduces the size of the spot market and the generation capacity that the dominant firm can offer in that market. We represent this effect in the right panel of the figure as a shift of the vertical axis. The forward position reduces the incremental revenue for each production level that the dominant firm would obtain by increasing the price. This happens because the contracted volumes are sold at a fixed price. As a consequence the spot-market price that maximises the dominant firm's profit is lower than if the contract were absent. This provides the basis for requiring the dominant generator to commit to forward sales. Such a measure was implemented, for example, in Alberta (Canada).[41]

Market power mitigation through financial contracts presents some advantages over asset divestiture. First, it does not risk causing technical inefficiencies by splitting the physical generation capacity among multiple firms. Second, it may be easier to implement than asset divestiture. Finally, measures based on financial contracts can be easily terminated, once the structural conditions driving the market-power concerns have been alleviated by entry in the market of other generators.

However, forward contracts can be less effective than structural measures. First, in a repeated context, the dominant firm may find it profitable to set spot prices higher than those maximising short-term profits, in order to influence the outcome of future sales of forward contracts. The price for the forward contracts will be higher if the buyers expect higher spot market prices during the delivery period. If the dominant firm could commit to the contracts' buyers to set the monopoly price in the spot market throughout the delivery period, the profits for the dominant firm from the forward sales would be the same as those achievable on the spot market without the mitigation measure. More generally, the dominant firm might find it optimal to forgo some spot market profits in order to influence the value of the forward contracts sold in the next round.

Second, as we discuss next in the context of virtual power plants, forward contracts yield spot market prices higher than those that would prevail if the same level of capacity was divested.

Virtual power plants

A virtual power plant (VPP) is an option contract allowing the holder to buy a certain volume of electricity at a predetermined price, known as the 'strike price', and resell it at the spot price. The option will then be exercised whenever the spot market price is higher than the strike price. In other words, a VPP has the same effect as a forward contract when the spot-market price is above the strike price, and produces no effect when the spot-market price is below the strike price.

The VPP reduces the incentives for the dominant firm to exercise market power, by making the profit on the contracted quantity independent of the spot price. Conversely, when the spot price is below the strike price the VPP has no impact on the firm's profit.

Note that the forward contracts analysed in the previous section can be interpreted as VPP with a zero strike price, which will therefore always be exercised. Therefore, other things being equal, a VPP will have a market-power mitigation effect that is at most equivalent to that of a forward contract for the same volume.

Federico and López (2009) compare the effect on the spot price of the transfer of a set of generation assets with that of VPPs hedging the same assets.[42] They find that the VPPs have a milder market-power mitigation effect than divestiture. This happens because, contrary to divestiture, VPPs do not affect the production capacity available to the competitors. As a comparison between Figures 5.7 and 5.8 reveals, both measures cause a shift in the residual demand curve faced by the dominant firm, but only the divestiture changes its slope, by removing from the dominant firm the right to set the offer price in the spot market for the divested capacity.

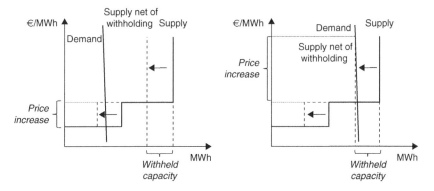

Figure 5.9   Effect of capacity withholding on market price

5.4.2   Price Caps

An overall price cap is implemented in most electricity spot markets, at least in the form of an administratively set price applicable to the transactions that take place under conditions of scarcity, when there is physical rationing of demand.

Setting a price cap below the VoLL may effectively mitigate market power, because spot electricity markets are particularly vulnerable to market power when demand approaches the available generation capacity. Since when the system is tight both demand and supply are price inflexible, withholding even a small amount of generation capacity from the market may result in a dramatic price increase. This holds especially if withholding capacity causes the market-clearing price to rise above the system marginal cost.

Figure 5.9 compares the impact of capacity withholding under contrasting conditions. In the left panel demand is low relative to the available generation capacity; capacity withholding therefore moves the market-clearing price along the system variable cost, to the variable cost of a more expensive generator. In the right panel the existing generation capacity is close to full utilisation; in this case capacity withholding induces scarcity, and the market-clearing price moves from the variable cost of the most expensive generator to the much higher level necessary to ration demand.[43]

As the figure shows, when the system is tight even relatively small generators may have the incentive and the ability to set extremely high prices, because all it takes to bring about a large price increase is a small reduction in supply.

Ruling out the possibility that the market-clearing price departs significantly from the system, marginal cost reduces the incentives to exercise market power. In this case the price increase caused by capacity withholding is relatively limited, even if the system is tight.

However, setting an overall price cap lower than the VoLL also prevents the price reaching the efficient level in genuine conditions of scarcity, making it impossible to remunerate the capital invested in (the efficient level of) generation capacity. For this reason, capacity support schemes are generally implemented together with price caps.[44]

Note, finally, that an overall price cap may lead to inefficiency in the event that demand shows some elasticity at a price above the cap. In this case the cap prevents the market-clearing price from rising to the level that rations demand in situations of scarcity, increasing the need for physical rationing.

5.4.3   Bid Mitigation

Spot-market operators in the US implement *ex ante* market-power mitigation mechanisms that force the bids of certain generators to predefined reference levels when certain conditions occur.[45]

The first stage of automatic mitigation procedures is a structural test. In some implementations the structural test directly triggers enforcement of the caps on offer prices. For example, the PJM Interconnection[46] tests whether any three suppliers are jointly pivotal for resolving any transmission constraints.[47] The three pivotal supplier assessment for a transmission constraint starts with a baseline scenario, that is, the generator dispatch and corresponding power flows on the network that would take place if the market were cleared without enforcing the relevant transmission constraint.[48] The baseline scenario is then used to check whether the constraint could be resolved in the event that all three pivotal suppliers refused to change their output from the base levels, by re-dispatching other units with incremental costs less than or equivalent to 150 per cent of the baseline market-clearing price. Finally PJM applies offer capping to the units that fail the three pivotal supplier test. This type of screening and bid mitigation is performed on both the day-ahead and real-time markets.

In other markets the structural test serves as a first screen to determine whether to subject specific suppliers to further tests, which may then trigger bid mitigation. The structural test is specifically used to identify areas of the network that are presumed to be prone to non-competitive outcomes due to transmission limitations. In competition policy terms, the structural test can be regarded as identifying the relevant market, a preliminary step to the competitive assessment carried out through additional tests. Two types of additional tests are then implemented. Conduct tests compare a unit's offer to a predefined threshold (for example, 300 per cent above the unit's cost-based reference price). Impact tests detect whether bids that failed the conduct test increase the clearing price above the level that would result if the reference prices were bid. Conduct and impact tests are carried out for all supply resources located in the system operator's control area. However, more stringent conduct and impact thresholds apply to generators located in the potentially non-competitive areas identified by the structural tests. If these tests are failed, bid mitigation is implemented.

The same conceptual framework is applied in the US capacity markets, with two additional features related to the specific nature of these markets The first is that the cap on the offers in the capacity markets reflects units' expected profits in the energy and operating reserve markets. Recall from Chapter 3 that capacity markets are a means to support generators' income, in the event that the prices prevailing in the energy and ancillary service markets are too low to attract an adequate level of investment. Therefore, the higher the income that a unit is expected to obtain by selling energy and operating reserve, the lower the reference price for that unit in the capacity market. The second feature is that entrants are generally regarded as competitive price setters and therefore not subject to the mitigation mechanism.

Besides generators' offer prices, other practices are monitored in some US markets. These include the physical withholding of generation resources, uneconomical production that cannot be justified and load bidding or virtual bidding practices that create unwarranted divergence between day-ahead and real-time prices. The physical withholding of capacity from the market has the same effect on the market price as offering it at prices above the market-clearing level. Uneconomical production, typically aiming at creating congestion, is detected by comparing the market-clearing price with the unit's reference cost level. Load-serving entities may also exercise (buyer) market power in the day-ahead market by consistently scheduling load in real time. Finally, virtual bids, meant to ensure the consistency of day-ahead and real-time prices, may be misused in order to induce divergences between day-ahead and real-time prices that are not justified by the fundamental market conditions and information available to market participants.

5.5   SELECTED CASES

In this section we discuss cases in which the technical features of electricity are particularly influential in shaping competition policy issues.

5.5.1   Market Power and Market Design: Bidding in Situations of Congestion

In most European countries forward electricity markets, including the day-ahead and the intraday markets, are run as if transmission capacity is unlimited. In the event that the power flows implementing the market outcome violate one or more network constraints, generators and possibly load-serving entities are paid to modify the level of production and consumption they scheduled after the unconstrained trading stage. Selection of the offers to modify the production and consumption programmes takes place in the real-time market or in an *ad hoc* re-dispatch market.

In Chapter 4, Section 4.4 we showed that this feature of the market design has an impact on the generators' profit-maximising offer strategy. Consider first, generators located in import-constrained areas, where production will have to be increased by the system operator at the re-dispatch stage in order to address the network constraint, and where the clearing price of the re-dispatch market is therefore higher than the clearing price of the forward markets. Generators located in the import-constrained areas will maximise their profits by withholding capacity from the unconstrained forward markets and offering it on the re-dispatch market, where its value is higher. Specifically, these generators have an incentive to offer their production on the forward markets at the price they expect will clear the re-dispatch market in the area where they are located.

A matching incentive holds in the export-constrained areas of the network, where production will have to be reduced by the system operator at the re-dispatch stage in order to address the network constraint, and where the clearing price of the re-dispatch market is therefore expected to be below the unconstrained clearing price of the forward markets. Generators located in export-constrained areas will maximise their profits by selling on the forward markets at any price higher than the expected clearing price of the re-dispatch market, irrespective of their variable costs.[49] By doing so, these generators will not produce, but they will cash in on the difference between the price obtained from their forward sales and the price that they pay to the system operator at the re-dispatch stage to reduce production.

As we argued in Chapter 4, Section 4.4.2, these incentives hold even for generators that do not have any market power. Nevertheless, generators' bidding behaviour in the event of congestion has been subject to scrutiny by the competition authorities as potential abuse of market power in some European markets. In Spain between 2002 and 2005 the competition authority brought a series of legal cases against generators whose bids in the day-ahead market reflected the expected price in the re-dispatch market.[50] The Spanish competition authority regarded the offer strategy as an abuse of dominant position.[51]

In the UK the energy regulator Ofgem, which is also in charge of enforcing competition law in the electricity industry, has been concerned that the increase in congestion costs on the interconnector between Scotland and England was caused by abusive bidding behaviour. Ofgem examined the situation in September and October 2007, and found evidence that Scottish generators submitted lower bids in the balancing market than comparable generators located in England during periods of congestion on the Anglo-Scottish boundary in the direction of England, that is, when Scotland was export constrained. The Scottish generators were also found to have submitted higher offers than comparable plants located in England when congestion was in the direction of Scotland, that is, when Scotland was import constrained. The generators' behaviour also included 'the apparent withholding of in-merit plants during import constraints despite positive spreads being available in the forward market'. Other periods were identified in which generators appeared to be running plants out of merit during export constraints.[52]

Ofgem concluded that the observed behaviour provided no basis for an abuse of dominance allegation within the standard competition policy framework. Interestingly though, Ofgem has sought an extension of its powers, allowing it to sanction this kind of behaviour within a different legal framework.[53]

In other contexts, notably Germany and recently Spain, the problem is considered a regulatory issue, and proposals are put forward to address it by regulating offer prices in the re-dispatch market.[54]

The issues discussed in this section are a consequence of the high level of product standardisation implemented in the European electricity spot markets. In the forward markets, electricity produced at all network locations is conventionally considered as the same product, and therefore exchanged at the same price, even if network constraints make it physically impossible to substitute production at one node with production at another. However, when it comes to physical delivery, this fiction cannot be maintained: the production programmes selected by the market participants at the unconstrained stage have to be modified if some network security constraints are not met. At this point, the price for electricity at each node is driven by the physical market fundamentals, that is, it reflects the real marginal cost of meeting the incremental consumption at that node.

Measures that prevent market participants' bids and offers from reflecting the impact of network constraints on the value of electricity can hardly be considered market-power mitigation measures. Rather they force (competitive) generators to behave irrationally, in order to mitigate the adverse impact on customers of the artificial geographical product standardisation implemented in the forward markets.

5.5.2   Re-dispatch Cost and Cross-border Capacity: The Swedish Interconnector Case

In most European countries, transmission congestion within national borders is dealt with via re-dispatch. Cross-border congestion, on the other hand, is generally managed by limiting the volume of exchanges that cause cross-border flows. This is achieved by allocating market participants a set of feasible transmission rights, the volume of which is jointly agreed by the relevant national system operators. The Nordic electricity market, comprising the territory of Norway, Sweden, Finland and Denmark, presents an interesting example of a mix of the two approaches.

Part of transmission congestion in the Nordic exchange is dealt with by dividing the area into market zones. The market participants make bids and offers that specify in which zone the electricity purchased and sold will be respectively produced and consumed. The market-clearing price differs across zones in the event of congestion. Until November 2011 the market zones were the national territories of Sweden and Finland. There are two zones in Denmark and four in Norway.[55] Congestion within each market zone is dealt with via re-dispatch by the national system operator.

In this context, national system operators may face a trade-off between the cost of domestic re-dispatch and the level of cross-border transmission capacity allocated to the market. It may be the case that by limiting the cross-border trades a system operator can reduce the need for domestic re-dispatch. This situation is illustrated in a highly stylised setting in Figure 5.10. The transmission networks of two neighbouring countries, A and B, include, respectively, two and one nodes. The transmission capacity of the lines and the supply and demand, which as usual we assume is

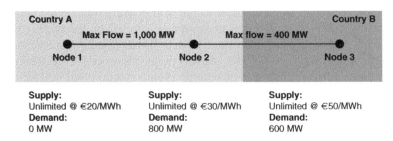

Country A		Country B
Max Flow = 1,000 MW	Max flow = 400 MW	
Node 1	Node 2	Node 3
Supply: Unlimited @ €20/MWh	**Supply:** Unlimited @ €30/MWh	**Supply:** Unlimited @ €50/MWh
Demand: 0 MW	**Demand:** 800 MW	**Demand:** 600 MW

Figure 5.10   Grid representation

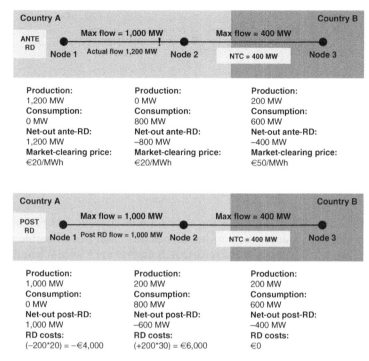

*Figure 5.11   Market outcomes with net transfer capacity (NTC) =
400 MW*

price inelastic, are indicated in the figure. Congestion between the two
countries is managed by allocating a feasible set of transmission rights to
the market. Congestion within each country is managed via re-dispatch.

Figure 5.11 illustrates the market outcome in the event that the system
operators decide to allocate the market 400 MW transmission rights
between country A and country B, the largest feasible set of transmission
rights. Under that assumption the market equilibrium is such that 400
MW of country B's demand and country A's entire demand are supplied
by generators connected at node 1, the cheapest producers. The remaining
200 MW demand in country B is supplied by domestic generators, connected at node 3. The market-clearing price is €20/MWh in country A,
equivalent to the unconstrained system marginal cost in the country, and
€50/MWh in country B.

The 1,200 MW flow between nodes 1 and 2 implementing this market
outcome is infeasible. This congestion, being internal to country A, is
managed through re-dispatch. Country A's system operator buys 200 MW

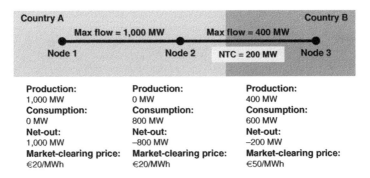

Figure 5.12   Market outcome with NTC = 200 MW

at node 2 at €30/MW and sells the same amount at node 1 at €20/MWh. This re-dispatch reduces the flow between nodes 1 and 2 to a feasible 1,000 MW. The clearing prices of the re-dispatch market are €30/MWh and €20/ MWh, respectively, at node 1 and node 2. The system operator bears the $200*(30–20) = €2,000$ re-dispatch cost.

Figure 5.12 shows the market outcome if only 200 MW of transmission rights between countries A and B were allocated to the market participants. In this case only 200 MW of country B's demand adds to country A's demand, which is met by the generators connected at node 1. The market outcome now results in a 1,000 MW power flow along the line connecting nodes 1 and 2. This flow is feasible, therefore no re-dispatch is required in country A. By limiting the volume of cross-border transmission rights, the system operator of country A therefore saves on re-dispatch costs. However, the total generation cost is not minimised.[56]

Furthermore, different levels of cross-border capacity lead to different distributions of the total surplus among the generators and consumers of the two countries, as well as the transmission right-holders.[57]

In April 2009 the European Commission opened a case against the Swedish system operator Svenska Kraftnät (SvK). The Commission was concerned that SvK was limiting the amount of export transmission capacity across Sweden's borders, with the objective of relieving transmission congestion on its internal network. According to the Commission, this would favour consumers in Sweden over consumers in neighbouring member states by reserving domestically produced electricity for domestic consumption.[58] The Commission's preliminary assessment found that this behaviour could represent an abuse of dominant position on the Swedish electricity transmission market, as it discriminated between domestic and export electricity transmission services.[59]

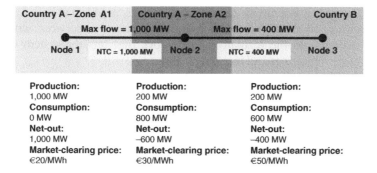

Country A – Zone A1	Country A – Zone A2		Country B
Max flow = 1,000 MW	Max flow = 400 MW		
Node 1	NTC = 1,000 MW	Node 2	NTC = 400 MW	Node 3

Production:	Production:	Production:
1,000 MW	200 MW	200 MW
Consumption:	**Consumption:**	**Consumption:**
0 MW	800 MW	600 MW
Net-out:	**Net-out:**	**Net-out:**
1,000 MW	−600 MW	−400 MW
Market-clearing price:	**Market-clearing price:**	**Market-clearing price:**
€20/MWh	€30/MWh	€50/MWh

Figure 5.13   Market output after the splitting of country A into two market zones

In order to address these concerns, SvK offered to subdivide the Swedish transmission system into two or more bidding zones, and to manage domestic congestion without limiting trading on the interconnectors.[60] In our example, this would amount to establishing two market zones within country A, as shown in Figure 5.13.

The system operator now allocates 1,000 MW transmission rights between the two market zones in country A and 400 MW between zone A2 and country B. The market outcome is shown in the figure. The demand faced by the cheapest generators located at node 1 is now limited to 1,000 MW by the available transmission rights. The second cheapest generators located in node 2 are activated to serve the remaining 200 MW demand in country A, limited by the available transmission rights from zone A2 to country B. Since both interfaces are congested, a different price clears the market in the each zone. No re-dispatch is necessary.

The solution proposed by the Swedish system operator eliminates discrimination, since cross-border and domestic transactions compete on a level playing field for the use of transmission resources. Note, incidentally, that the mechanism leads to the efficient allocation of transmission capacity and to minimisation of total supply costs. Therefore the system operator's proposal reconciles the non-discrimination objective and the efficiency objective, the standard reference for market design.

From an institutional perspective, in the Swedish Interconnector case competition policy has entered a territory traditionally within the domain of energy policy. By addressing the problem within the competition policy framework, the Commission has taken the regulatory system (and the regulators) out of the picture. The Commission sees limiting cross-border capacity in order to reduce re-dispatch cost as an abuse of dominant position under Article 102 of the Treaty. In so doing:

- the system operator's dominant position is established based on its monopoly position in the supply of transmission services, without reference to the regulatory framework that sets the system operator's objectives and incentives, and
- the abuse is characterized in terms of discrimination between domestic transactions and cross-border transactions without reference to efficiency issues, which would be the main focus of a regulatory approach.

5.5.3 Abuse of Dominant Position in Related Markets

Generally, wholesale spot prices are not directly passed on to final consumers through retail prices. Retail prices are typically fixed for a relatively long period, for example a year or a quarter, and reflect the wholesale prices expected when the retail contract is signed.[61]

Retailers hedge against wholesale price volatility by forward buying the volume of electricity that they expect to retail. Most retailers, however, are vertically integrated in the generation business, which provides a natural hedge against wholesale price volatility, by making them at the same time buyers and sellers on the wholesale market.

A congestion management system based on locational prices adds a geographical dimension to the hedging problem faced by the retailer. It needs to fix the price of the electricity in the market zone where its clients consume by purchasing forward contracts with delivery in that market zone. Alternatively, if the retailer owns generation capacity in a different zone, it can buy long-term transmission rights enabling it to 'move' electricity from the market zone where it can generate it to the market zone where its clients consume it.

The Italian electricity spot market began operating in 2004. At that time forward trading was limited and long-term transmission rights were not available. The main retailers were also generators. Congestion management was, and still is, carried out by splitting the Italian territory into market zones, the prices of which vary in case of congestion.

The peculiar feature of the Italian congestion management system is that locational prices are assessed for production only, while all consumption is charged a uniform nationwide price. The uniform national price (PUN) is calculated as the weighted average of the market-clearing prices of all the market zones, with weighting the equivalent of the demand in each market zone.[62]

Intuitively, this approach amounts to placing all consumers in a virtual market zone connected to the production zones in a way such that the clearing price in the virtual zone is the weighted average of clearing prices

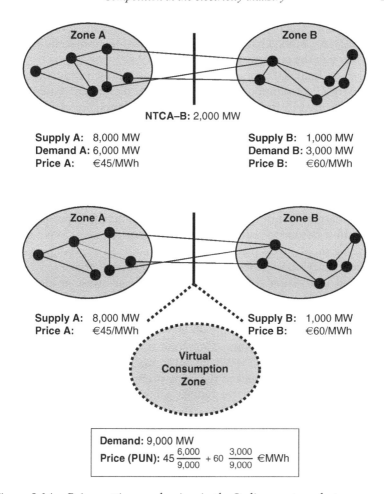

Figure 5.14   Price-setting mechanism in the Italian spot market

in the real zones, as illustrated in Figure 5.14 for the simplest two-zone market. The upper panel of the figure shows the market outcome resulting from the zonal market-clearing algorithm. The lower panel shows the market outcome obtained with the clearing algorithm implemented in Italy.

The transmission capacity between the production zones and the virtual consumption zone is allocated by a power exchange via an implicit auction, as illustrated in Chapter 4, Section 4.3.2. Bilateral transactions between a generator and a consumer attract a congestion fee equivalent to the difference between the uniform national price and the market-clearing price in the zone where production takes place. This congestion fee makes it indifferent for market participants to implement the transaction bilaterally or through the power exchange.[63]

Long-term financial transmission rights between the production zones and the virtual consumption zone are issued by the Italian system operator and allocated to the market. However, for simplicity of exposition, we illustrate the competition case with reference to a bilateral transaction between a supplier and its customers.

We consider a supplier retailing 90 MW to end-consumers under fixed-price annual contracts. We assume for simplicity's sake that no forward wholesale contracting takes place in the market and that the supplier controls some generation capacity. It is easy to verify that, in order to lock in the retail margin, that is, to make its profit independent of spot prices, the supplier must be able to produce 60 MW[64] in zone A and 30 MW[65] in zone B. Since the supply takes place via a bilateral contract the spot prices only enter the retailer's profit function via the congestion charge.

The unit congestion fee for a bilateral contract with production in zone A is $(PUN - PA)$, irrespective of where consumption takes place. For a bilateral contract with production in zone B the congestion fee is $(PUN - PB)$. Hence by producing 60 MW in zone A and 30 MW in zone B the retailer pays an average congestion fee of $60/90*(PUN - PA) + 30/90*(PUN - PB) = 0$ €/MWh. In other words, by allocating production across the market zones in the same proportion as the weighting of the zonal prices in the PUN, the retailer gains a perfect hedge against the volatility of spot market prices.

Consider now a retailer owning generation capacity only in zone A.[66] By producing the entire 90 MW supplied to its customers in zone A, the retailer pays unit congestion charge $PUN - PA = 1/3*(PB - PA)$. In this case the congestion fee paid by the retailer to supply its clients depends on the locational prices. If the spot prices turn out to be different from those expected at the time of committing to the fixed supply price, and especially if the difference between the spot prices in the different locations was wrongly anticipated, the retailer's profits will depart from the expected level.

In April 2005 the Italian competition authority started proceedings against Enel S.p.A., the main generator and retailer in the Italian market.[67] The competition authority identified some features of the company's bidding behaviour in the spot market that did not appear to reflect genuine profit maximisation. In the regulator's interpretation Enel's strategy reflected an aim of imposing some form of control over its competitors.[68]

The competition authority specifically regarded selective increase of the market price in the zone where Enel's market share in generation was largest as potentially abusive. The idea was that, by increasing the price in that zone, Enel's competitors that (a) did not own generation capacity in that market zone, and (b) had taken fixed price commitments in the retail market, would suffer a reduction in profits as a consequence of the increase in the congestion fee.

In our example the allegedly abusive behaviour consists of increasing the spot price in market zone B, which, other things being equal, would increase the congestion fee paid by a retailer that matched its sales to final consumers with production in zone A.

The case was dropped after the Italian competition authority accepted Enel's commitments to offer a certain number of forward contracts for delivery of electricity into the market zones where its market share was largest. In the context of our example, Enel committed to sell forward contracts for delivery at zone B, which would also allow retailers with no generation capacity in that area to hedge against the volatility of the congestion fee.

5.5.4 Abuse of Dominant Position in the System Operation Business

The independence of the system operator is generally recognised as a necessary condition for the successful liberalisation of electricity markets. System operators control access to the transmission network in the short run by deciding the volume of transmission rights made available to market and running the allocating mechanism, and in the long run by planning network development. Discrimination in network access and delays in network upgrades can hinder competition in the wholesale as well the retail markets. Furthermore, the system operator is also the only buyer of ancillary services, a major source of income for some generating units.

In most markets some form of unbundling of system operator activities has been implemented in order to ensure the system operator's independence. The solutions implemented differ in three broad dimensions. The first is the scope of the system operator's responsibilities. At one extreme the system operator is responsible only for real-time system balancing. At the opposite extreme it controls a wide range of activities, including the transmission asset maintenance schedule, network development planning and generation capacity adequacy.

The second dimension relates to the degree of unbundling of system operations from the generation business. At one extreme the system operator is a fully ownership-separated company. The alternative approach is based on ring-fencing system operator activities from the integrated owner's generation activities. This is pursued through measures such as fully separated management teams, separate physical premises, Chinese walls, and independent audit boards or steering committees including representatives of both stakeholders and the regulator.

Finally, system operation arrangements differ in the degree of integration between the system operator and the transmission network operator. In some cases the system operator owns the transmission network; in others the transmission network is owned and operated by a different firm.

In Europe, several member states have selected ownership unbundling of (integrated) system and transmission operators. These are the UK, Italy, several northern European countries and Spain.[69] The other countries have opted for functionally unbundled system operators. In the US, ownership and functionally unbundled system operators also co-exist.

Ensuring the system operator's independence is a crucial energy policy objective. However, in a recent case the organisation of system operations was shaped by competition policy. In 2008 the European Commission opened a case against E.ON AG,[70] at that time a large generator as well as the owner of a portion of the German electricity grid and the system operator of the area covered by its network. The concern was that E.ON might have abused its dominant position on the market for secondary reserve demand in the E.ON network area by discriminating between secondary reserve suppliers, favouring their affiliate generators, and by preventing generators from other member states from exporting balancing energy into the E.ON balancing market.

The case was dropped after the Commission accepted E.ON's commitments to divest its 380/220 kV-line transmission network, the system operation of the E.ON control area and related activities.

NOTES

1. Specifically, Article 102 of the Treaty on the Functioning of the European Union.
2. Bundeswettbetwerbhorde website, 12 June 2012.
3. For each price level, the residual demand represents market demand net of the volumes supplied by the firm's competitors. It summarises the effects of competition on the demand faced by the firm.
4. The European Commission in Case COMP/M.5549 EDF/SEGEBEL stated: 'It appears from the market investigation that a significant interaction exists between OTC traded electricity products and electricity products traded on organised markets. Similarly, sufficient interaction exits between financial and physical products as the former use the latter as underlying products'. See also European Commission's decisions in Case COMP/M.5224 – EDF/BRITISH ENERGY and Case COMP/M.5911 TENNET/ELIA/GASUNIE/APX-ENDEX.
5. The assumption that competitive conditions in the wholesale electricity market are similar across many hours is justified by the stability of the supply conditions. The dynamics of installed generation and transmission capacity unfold over timeframes in

the order of years, sometimes decades. Furthermore, as discussed in Chapter 2, generators' dynamic constraints make production and offer decisions in consecutive hours interdependent.

6. This assumes that congestion between the two zones is managed through locational pricing. For the discussion of competitive constraints in cases where transmission constraints are managed through re-dispatch, see Chapter 4.

7. See, for example, in Motta, M., 2004. *Competition Policy: Theory and Practice*, Cambridge: Cambridge University Press.

8. Commission Notice on the definition of the relevant market for the purposes of Community competition law (published in the *Official Journal*: OJ C 372, on 9 December 1997).

9. With the possible exception of very long-term multi-year contracts.

10. See European Commission's decisions in Case COMP/M.4180 – Gaz de France/ Suez; Case COMP/M.5224 – EDF/BRITISH ENERGY; Case COMP/M.5549 EDF/ SEGEBEL; and Case COMP/M.5911 TENNET/ELIA/GASUNIE/APX-ENDEX.

11. In Case COMP/M.3868 DONG/Elsam/Energi E2, for example, the Commission assessed that balancing energy and ancillary services are not easily substitutable by other electricity supplies. However, more recently, in Case COMP/M.5224 EDF/ BRITISH ENERGY, the Commission found that the British balancing market is not separate from the forward markets.

12. In US jargon, those generators apply for 'market-based rate authority'. See AEP Power Marketing, Inc. et al., 107 FERC 61,018 (2004).

13. Ibid., p.43.

14. Nine periods are obtained by aggregating the off-peak, shoulder and peak hours in winter, summer and spring/autumn. The tenth period includes summer super-peak hours.

15. The European Commission has made no explicit assessment regarding separation of the markets depending on the time of delivery. In Case COMP/M.5467 RWE/ESSENT the European Commission stated: 'Within the market for generation and wholesale supply, the Dutch Competition Authority (the NMa) distinguishes between peak hours and off-peak hours. In addition, they also consider the possibility of a further distinction between peak and super-peak hours. However the response to the current market investigation in this regard was inconclusive. The definition of the relevant product market can therefore be left open, as this does not change the final assessment'.

16. Moselle, B., Newbery, D. and Harris, D., 2006. *Factors Affecting Geographic Market Definition and Merger Control for the Dutch Electricity Sector*, The Brattle Group Limited. Report undertaken for the Dutch Competition Authority (NMa).

17. 'Persistent differences in average peak prices and average off-peak prices indicate that peak and off-peak power may be separate products. If they were not, then presumably consumers would substitute peak and off-peak power until the prices of the two products came closer to one another' (ibid., p.15).

18. The European Commission's market investigations in the context of merger control found that the geographical markets are normally national, but that they may sometimes be smaller or larger. Relevant elements considered in the Commission's analysis include market design, the existence of congestion and the existence of price correlations and price differentials across locations.

   In Case COMP/M.3268 Sykdraft/Graninge, the Commission investigated the frequency and distribution of the different price areas in the Nordic electricity market. In Case COMP/M.3868 DONG/Elsam/Energi E2, the Commission ruled out a Nordic-wide market definition, based on the assessment that for a very substantial fraction of the hours producers in West Denmark do not suffer any competitive constraints from producers in the rest of the Nordic region.

19. O'Donoghue, R. and Padilla, A.J., 2006. *The Law and Economics of Article 82 EC*, Oxford: Hart.

20. The HHI index ranges from 0 to 10,000.

21. In the Cournot model all suppliers simultaneously select the volume offered on the market.
22. Borenstein, S., Bushnell, J. and Knittel, C., 1999. 'Market Power in Electricity Markets: Beyond Concentration Measures', *Energy Journal*, **20**(4), 65–88.
23. Fabra, N., de Frutos, M.-A. and von der Fehr, N-H., 2008. 'Investment Incentives and Auction Design in Electricity Markets', CEPR Discussion Papers, London.
24. An example can be found in Green, R. and Newbery, D.M., 1992. 'Competition in the British Electricity Spot Market', *Journal of Political Economy*, **100**(5), 929–53.
25. Wolak, F., 2000. 'An Empirical Analysis of the Impact of Hedge Contracts on Bidding Behaviour in a Competitive Electricity Market', *International Economic Journal*, **14**(2), 1–40; Wolak, F., 2003. 'Measuring Unilateral Market Power in Wholesale Electricity Markets: The California Market, 1998–2000', *American Economic Review*, **93**(2), 425–30; Baselice, R., 2007. 'Italian Power Exchange and Unilateral Market Power in Italian Wholesale Electricity Market', paper presented at the 9th IAEE European Energy Conference, Florence, Italy, 10–13 June.
26. Below the value of lost load (see Chapter 2, Section 2.2.1).
27. Newbery, D.M., Green, R., Neuhoff, K. and Twomey, P., 2004. *A Review of the Monitoring of Market Power*, ETSO Report, November.
28. In a joint market investigation the Italian energy and competition authorities developed an index reflecting the possibility and the incentives for generators to exercise market power. The index is based on a comparison of the outcome for the generator of two strategies: selling pivotal capacity at the maximum possible price (for example, the price ceiling of €500/MWh implemented on the Italian power exchange), or selling all its production at the estimated system marginal cost ('Fact-finding investigation into the state of liberalisation in the electricity sector', 2005, AGCM and AEEG).
29. If the applicant fails one or both tests, it can rebut the presumption of market power by offering additional evidence.
30. See Section 5.4.3.
31. As reported by Newbery et al., 2004 (see n. 27, above), p. 28. Sheffrin, A., 2002. 'Predicting Market Power Using the Residual Supply Index', paper presented at the FERC Market Monitoring Workshop, Washington, DC, 3–4 December.
32. Italian competition authority (AGCM) jointly with the Italian energy regulator (AEEG), 2005. *Fact-finding Investigation into the State of Liberalisation in the Electricity Sector (IC22)*.
33. Congestion management in the Italian market is carried out by allocating the market transmission rights between market zones. As a consequence the electricity market-clearing prices may be different in different zones. A macro zone is an aggregate of neighbouring market zones in which the market-clearing price is often identical. AGCM carried out the pivotality assessment by macro zones for reasons of simplicity.
34. Borenstein, S., Bushnel, J. and Wolak, F., 2002. 'Measuring Market Inefficiencies in California's Restructured Wholesale Electricity Market', *American Economic Review*, **92**(5), 1376–405; Joskow, P. and Kahn, E., 2002. 'A Quantitative Analysis of Pricing Behaviour in California's Wholesale Electricity Market During Summer 2000', *Energy Journal*, **23**(4), 1–35; Mansur, E., 2001. 'Pricing Behavior in the Initial Summer of the Restructured PJM Wholesale Electricity Market', POWER Working Paper 083; *Sector Inquiry under Art 17 Regulation 1/2003 on the gas and electricity markets (final report)*, 2007, COM(2006) 851 final; Weigt, H. and Von Hirschhausen, C., 2007. 'Price Formation and Market Power in the German Electricity Wholesale Market – Is Big Really Beautiful?', Electricity Markets Working Papers WP-EM-16, Dresden University of Technology.
35. More details on the issue of assessing generating costs are given in Harvey, S. and Hogan, W., 2001. *Identifying the Exercise of Market Power in California*, LECG, LLC, Cambridge, MA; Harvey, S. and Hogan, W., 2002. *Market Power and Market Simulations*, LECG, LLC, Cambridge, MA; Smeers, Y., 2005. 'How Well Can One

Measure Market Power in Restructured Electricity Systems?', CORE Discussion Paper No. 2005/50; Rajaraman, R. and Alvarado, F., 2003. 'Disproving Market Power', PSERC Working Paper; and Mansur, E., 2008. 'Measuring Welfare in Restructured Electricity Markets', *Review of Economics and Statistics*, **90**(2), 369–86.

36. In this section we draw extensively from Federico, G. and López, A., 2009. 'Divesting Power', IESE Business School Working Paper No. 812.

37. Examples of mergers or joint ventures in the electricity sector where divestments or virtual power plants (VPPs) have been required by the competition authorities include Gas Natural/Union Fenosa (2009), EDF/British Energy (2008), Gas Natural/Endesa (2006), GDF/Suez (2006), Nuon/Reliant (2003), ESB/Statoil (2002) and EDF/EnBW (2000). Alleged abuse of dominance cases where divestments of generation capacity or VPPs have been implemented as a remedy include proceedings involving E.ON (2008), RWE (2008) and Enel (2006). Divestments of power plants have also been used by regulators to mitigate the market power of incumbent generators in the UK and Italy in the 1990s, while in Spain and Portugal regulatory contracts and more recently VPPs have been employed to make the electricity market more competitive.

38. Federico and López 2009 (see n. 36, above).

39. This result provides a foundation for the widely held view that ownership of price-setting units grants greater market power than ownership of base-load capacity.

40. Some relevant references are: Allaz, B. and Vila, J.-L., 1993. 'Cournot Competition, Forward Markets and Efficiency', *Journal of Economic Theory*, **59**, 1–16; Newbery, D., 1998. 'Competition, Contracts and Entry in the Electricity Spot Markets', *RAND Journal of Economics*, **29**, 726–49; Green, R., 1999. 'The Electricity Contract Market in England and Wales', *Journal of Industrial Economics*, **47**, 107–24; and Bushnell, J., 2007. 'Oligopoly Equilibria in Electricity Contract Markets', *Journal of Regulatory Economics*, **32**, 225–45.

41. In 2000 the physical rights to output of existing generators built under the previous regulatory environment were auctioned off as Power Purchase Arrangements (PPAs). PPAs are auction biddable contracts that cover the embedded costs of existing generation. The PPAs were auctioned to interested parties. The holder of a PPA is entitled to sell the output of the generating plants directly to consumers in exchange for paying the owner the actual cost of generating power over the remaining life of the facility or until 2020, whichever comes first.

42. These are a series of options, each with strike price equal to the variable cost of the generating unit that it is meant to hedge.

43. A condition that we analysed in Chapter 2, Section 2.2.1.

44. See Chapter 3.

45. For a detailed analysis of the market-power mitigation mechanisms implemented in the US, see Reitzes, J., Pfeifenberger, J., Fox-Penner, P., Basheda, G., Garcia, J., Newell, S. and Schumacher, A., 2007. 'Review of PJM's Market Power Mitigation Practices in Comparison to Other Organized Electricity Markets', Brattle Group, Cambridge, MA, September.

46. PJM Interconnection is a US Regional Transmission Organisation (RTO) that coordinates the movement of wholesale electricity in all or parts of 13 states and the District of Columbia.

47. By assessing whether any three non-affiliated suppliers together are pivotal, PJM mitigates the unilateral and also the coordinated exercise of market power.

48. Recall that in the US standard market design transmission constraints are enforced within the market-clearing algorithm. Therefore market-clearing dispatch in the baseline scenario meets all the network constraints but the one for which the three pivotal supplier test is being carried out.

49. This might lead the generators to offer prices lower than their variable costs.

50. Tribunal de Defensa de la Competencia: Resolución Expte. 552/02, Empresas eléctricas; Resolución Expte. 602/05, Viesgo Generación; Resolución Expte. 601/05, Iberdrola Castellón and Resolución Expte. 624/07, Iberdrola.

51. For example in: Resolución Expte. 602/05, Viesgo Generación the Spanish competition authority states: '*ENEL VIESGO GENERACIÓN, S.L., es responsable de una infrac-ción al artículo 6 de la LDC, consistente en abusar de su posición de dominio en el mercado de energía eléctrica en una situación de restricciones técnicas de las zonas Centro-Sur y Sur, los . . . ofertando al mercado diario a precios superiores a sus Costes Variables Revelados, con el objeto de no casar en el mercado diario y sabiendo que sería llamada a restricciones técnicas, y pagada a su precio de oferta al diario, porque su energía era necesaria para satisfacer la demanda de la zona, al ser la única disponible en la misma*'.

52. Ofgem, 2009, *Addressing Market Power Concerns in the Electricity Wholesale Sector – Initial Policy Proposals.*

53. UK Energy Act 2010, Chapter 27, Part 3.

54. For Spain, see Comisión Nacional de Energía: *Propuesta de retribución regulada para el mecanismo de resolución de restricciones técnicas*, 15 April 2010, avail-able at: http://www.cne.es/cne/doc/publicaciones/cne27_10.pdf. For Germany, see Bundesnetzagentur *Redispatch-Workshop bei der Bundesnetzagentur am 07.12.2011, Vergütung für Redispatchmaßnahmen*, available at: http://www.bundesnetzagentur.de.

55. See Elspot market overview, available at: http://www.nordpoolspot.com/Market-data1/Maps/Elspot-Market-Overview/Elspot-Prices/.

56. It is easy to verify that the total generation cost is lower when the interconnection capacity is set at 400 MW and re-dispatch takes place.

57. The different distribution effect is caused by the different allocation of production among generators at the various locations and at the different market-clearing prices. The latter effect does not show in our simple example with flat supply functions.

58. Case/COMP 39.351, *Commission Opens Proceedings Against Swedish Electricity Transmission System Operator Concerning Limiting Interconnector Capacity for Electricity Exports*, MEMO/09/191, 23 April 2009.

59. Case/COMP 39.351, *Commission Market Tests Commitments Proposed by Svenska Kraftnät Concerning Swedish Electricity Transmission Market*, IP/09/1425, 6 October 2009.

60. Case/COMP 39.351, *Commitments under Article 9 of Council Regulation No 1/2003*, Svenska Kraftnät, 1 September 2009.

61. In some cases, retail prices are indexed to the value of a basket of fuels that is meant to reflect the dynamics of the (incremental) cost of generation in the market. However, this kind of indexation has no impact on the issues discussed in this section.

62. Congestion management via locational prices implies that consumers pay different prices for electricity solely because of their location. However, nationwide price uni-formity for essential goods such as electricity has traditionally been a social policy objective in Italy. The asymmetric pricing system was introduced to overcome political opposition to locational prices. As long as demand is price inelastic, the absence of locational signals on the demand side does not produce inefficiencies, because charging loads the uniform price or the nodal price does not change withdrawals. Inefficiency caused by lack of locational signals on the demand side is more likely to result in a longer-term horizon, to the extent that (especially) large customers' location decisions are affected.

63. Further detail on the congestion management system implemented in the Italian market can be found in the Vademecum of the Italian Power Exchange, available at: http://www.mercatoelettrico.org/En/MenuBiblioteca/Documenti/20091112VademecumofIpex.pdf.

64. Obtained as (6,000/9,000)*90MW.

65. Obtained as (3,000/9,000)*90MW.

66. And that cannot enter a forward contract for delivery in zone B.

67. Autorità Garante per la Concorrenza e il Mercato: Opening of Proceedings No. 14174 – A366, *Comportamenti restrittivi sulla borsa elettrica*, 6 April 2005.

68. In the Opening of Proceedings (ibid.) the Italian Competition Authority states: '*La strategia di Enel, nella sua globalità, appare riconducibile ad un'unica finalità, che è*

quella di esercitare il proprio potere di mercato dettando le strategie di prezzo all'ingrosso dell'energia elettrica e determinando, oltre le proprie, anche le condizioni concorrenziali dei suoi concorrenti, in modo tale da evitare confronti competitivi anche in quei mercati geografici rilevanti ove ciò avrebbe potuto, almeno potenzialmente e limitatamente, avvenire (il Nord e la Sardegna)'.

69. See Stern, J., 2011. 'System operators: lessons from US and EU energy industry experience and implications for the England and Wales water industry', CCRP Working Paper No. 18.
70. Case COMP/39.389 – German Electricity Balancing Market, 26 November 2011.

6. Retail competition

Anna Creti and Clara Poletti

6.1 INTRODUCTION

At the initial stages of the liberalisation process, most of the design and regulatory effort was typically focused on creating efficient wholesale electricity markets. In most European countries, and many jurisdictions in the US, retail markets are still in their infancy.[1] In several jurisdictions of the United States the incumbent utility still acts as the monopoly supplier for small consumers, procuring electricity in the wholesale market and retailing it at regulated prices. In Europe, despite the introduction into law of retail competition, in some countries entry barriers have not yet been completely removed and retail price controls are still enforced.

A dynamic retail market may be beneficial to the wholesale market, increasing the responsiveness of demand to price and contributing to the development of the long-term forward electricity market.

Since retail costs account for only a small share of the total cost of the electricity service, the potential for retail competition to lower consumers' bills by reducing retail costs is limited. Therefore, most of the benefits of retail liberalisation are related to the ability of competitive retailers to provide services that are tailored to consumer preferences.

Large consumers have diverse needs of energy price certainty, sophisticated procurement strategies, and in some cases flexibility in the use of electricity. They are therefore in a position to take full advantage of retail competition. However, the value of a wider range of offers for small consumers is less evident at this stage.

Smaller consumers appear to face significant transaction costs in order to identify, assess the offers of and switch to a different supplier. As a consequence, the incumbent retailer enjoys significant market power over its passive customers. This has led regulators in most jurisdictions to retain price controls long after the legal liberalisation of electricity retailing. Even in the UK, where the electricity retail market is far more developed than in most other European countries, six years after all price controls were lifted, a prohibition of undue discrimination between residential customers has been introduced, and further regulatory constraints for retailer pricing strategies are currently being discussed.

Only small numbers of consumers switch to a competitive retailer in markets where regulated tariffs are available. This is indicative of a complex trade-off between the protection of passive consumers against the incumbent's market power, and the development of competition in electricity retailing. No clearly superior set of policies addressing this trade-off has emerged so far.

In Section 6.2 we describe electricity retail activity and its typical risk structure. In Section 6.3 we discuss the nature and scope of the benefits of retail competition for different types of consumers. Finally, in Section 6.4 we investigate the interaction between competition and regulation in electricity retailing.

6.2 ELECTRICITY RETAILING

Like retailers in other sectors, electricity retailers bundle the inputs necessary to serve final electricity consumers. Specifically, electricity retailers:

- are responsible for procuring from the wholesale market the electricity consumed by their clients, and are liable to imbalance charges on the difference between their clients' consumption and the volume procured (see Chapter 2, Section 2.3.1);
- procure system operation, transmission, distribution and metering[2] services;
- design and advertise offers appealing to the different types of consumers;
- act as an interface for their clients on matters related to the electricity service; and
- issue invoices and collect payments.

The typical risk structure of electricity retailers can be outlined as follows. Electricity retailers are best positioned to take on the commercial risk, given their superior knowledge of their clients' creditworthiness. Instead, the energy price risk is typically hedged by retailers. Most small consumers are supplied under fixed-price contracts over a period of up to one or two years. In order to eliminate the possibility of gaps between the revenues collected from their fixed-price clients and the procurement cost, electricity retailers purchase electricity forward or buy financial hedges against the volatility of the spot price.

Transmission and distribution tariffs are typically passed on to the consumers,[3] as these costs are regulated, and therefore beyond the suppliers' control.

Industrial and large commercial consumers, which can accurately predict their consumption, generally agree with the retailer on the price for consumption notified at gate closure of the wholesale market.[4] In this case the supplier is responsible for procuring the notified volume from the wholesale market, while any imbalance costs are passed on to the consumers, so that the consumers bear the risk that the imbalance price might be different from the supply prices agreed with the retailer for the nominated quantity. In contrast, small consumers are typically offered a price for their actual consumption, with no *ex ante* quantity commitment. In this case the imbalance price risk is typically borne directly by the retailer, as the party better positioned to forecast its clients' withdrawals.[5]

6.3   COSTS AND BENEFITS OF RETAIL ELECTRICITY LIBERALISATION

We shall now discuss the potential benefits and costs of the introduction of competition in electricity retailing.

6.3.1   Product Choice and Service Quality

Retail activity usually creates value for consumers by providing information about the features, quality and price of the alternative products, by making products available at convenient locations and by providing before- and after-sales customer service.

The special technical features of electricity limit the scope for creating value in some of these areas. First, electricity cannot be displayed, so retailers cannot differentiate on how they present the product to the consumers. Second, electricity is carried directly to consumers' premises via the distribution network. Therefore retailers cannot differentiate how they deliver the service or the technical attributes of the electricity that their clients receive. Finally, product returns do not occur with electricity.

However, competing retailers can create value in two important areas: the design of the energy products and the quality of their customer service. Regulated utilities generally segment markets by broad customer classes, for example, residential, commercial and industrial. Within each class, further segmentation is sometimes based on the size of the consumers. The objective of capturing different consumer preferences does not appear to be central in the traditional regulated rate design. Furthermore, in traditional tariff setting, fairness and income-distribution concerns may be more influential than the objective of conveying the correct economic signals to consumers.

Competitive retail markets are expected to deliver more cost-reflective prices and greater product differentiation. The main differentiation between electricity products lies in the degree of price stability. Commonly observed pricing options range from fixed-price options for up to a few years to day-ahead or real-time prices. In some cases, retail prices are linked to indices that are intended to follow the dynamics of generation cost, such as gas price indices.

An increasingly important element of product differentiation is the environmental content of the offering. Popular options for the mass market include CO_2 offsets and renewable guarantees of origin. More complex packages are offered to businesses, enabling customers to diversify their sustainability portfolio.

Furthermore, retailers can add value over and above the simple supply of electricity by providing a number of related services ranging from the engineering and operation of energy facilities and energy management systems, to energy portfolio management, risk management, consulting and strategic energy sourcing.

Finally, retailers can differentiate in areas such as billing services, customer care and flexibility, as well as in attributes such as their social commitment and reputation for reliability.

6.3.2 Retailing Costs and Wholesale Procurement

Retailing accounts for a small part of overall supply costs, typically around 3–5 per cent. Therefore the scope for competition to reduce retail costs appears limited. A 25 per cent reduction in retail costs entirely passed on to the consumers would turn into a 1 per cent reduction of the bill for a typical residential consumer, a negligible amount in terms of monetary savings. In addition, retail liberalisation may lead to an increase in retail costs, especially in advertising, customer management systems (and their duplication) and transaction costs relating to the supplier switching process.[6]

Most electricity retail markets are highly concentrated. In 2009 in seven European countries the three largest suppliers' market share stood at 100 per cent or very little below. Only in four countries was the share below 50 per cent. The remaining 10 countries had total market share for their three-largest suppliers of between 58 and 97 per cent.

However, there is no compelling evidence that high levels of concentration in electricity retailing result in significant market power. In 2007 in Norway, for example – 10 years after liberalisation of retailing to small consumers became effective – the market share of the larger retailer in each area was on average above 70 per cent, but the retail margins were small and the prices offered by the main retailers systematically close to spot wholesale prices, signalling intense competition.[7]

Finally, some argue that competitive retailers may be more effective buyers on the wholesale electricity markets than a regulated monopoly retailer, thanks to superior analysis and negotiation practices.[8] Underlying this assessment is the idea that competitive retailers have greater incentives to minimize procurement costs than a regulated monopoly that is allowed to pass on its procurement costs to its franchised customers. This holds all the more if the monopoly retailer is also a generator, in which case high procurement costs benefit the company's generation business, without hurting its retail business. However, properly designed incentive-based regulation may provide strong incentives to the monopolist to minimize procurement cost. Furthermore, the large and captive customer base could place a monopoly retailer in a stronger bargaining position on the wholesale market and lead to lower procurement costs, especially if the wholesale market is less than competitive.

6.3.3   Wholesale Market Efficiency

A lively retail market may be beneficial to the wholesale market. In the previous section we mentioned the possibility that retailers could be more effective buyers in the wholesale market, compared to a regulated monopoly retailer. In addition, competitive retailers could contribute to the development and exploitation of demand-response capabilities. This would make the wholesale market less vulnerable to market power, reduce the need to resort to administratively set prices when the generation capacity is insufficient to meet demand, and possibly reduce the level of generation and transmission capacity needed. However, if consumers are load profiled, retail competition could lead to prices differing from the socially optimal levels a monopoly retailer would be able to charge.[9]

Retail competition is a necessary complement to the liberalisation of power generation for the development of the long-term contract market. Under retail liberalisation, consumers can express, through their retailers, their demand for long-term price certainty and exchange a hedge against spot-price volatility with the generators. If the retail market is not liberalised, the demand for long-term contracts from the monopoly retailer is likely to be shaped around the rules on cost pass-through set by the regulator. For example, if the regulator allows the wholesale spot prices to be fully passed on to retail tariffs, the monopoly retailer has little incentive to buy forward in order to hedge against spot price volatility. In this case the main source of hedging for the generators would be missing.

Finally, the largest electricity retailers are typically vertically integrated in the generation business. This suggests that the relationship between wholesale and retail markets is bi-directional. An imperfect wholesale market may hinder the development of retail competition, by increasing the cost to non-integrated retailers of hedging against the variability of wholesale electricity prices, by far their most expensive input. In the UK, for example, the regulator's concern that the wholesale products currently available do not meet the risk management needs of the independent suppliers (and generators) has led to a stream of policies aimed at increasing liquidity in the wholesale market.[10]

6.3.4   Energy Conservation and Distributed Generation

Provided that the appropriate incentives are in place, competitive retailers may contribute to reducing the implementation costs of sustainability policies, for example by encouraging consumers' participation in energy efficiency programmes and the development of on-site renewable generation.

Retailers could, for example, offer bundles including the supply of electricity and possibly gas, the hosting by consumers of photovoltaic generators owned and run by the retailer, load management and possibly direct load control, and administration of consumers' participation in publicly funded energy-saving schemes.

6.4   RETAIL COMPETITION AND REGULATION

Small European electricity consumers – and among those vulnerable consumers especially – do not yet appear to fully benefit from retail competition. Despite some positive developments in several European countries, switching rates remain low in many markets, as shown in Figure 6.1. Across the EU member states, only 10.1 per cent of household consumers enquired about switching supplier in the two years prior to 2010, and 6.2 per cent actually switched supplier.[11]

Small consumers appear to face significant transaction costs in order to identify, assess the offers of and switch to different suppliers. This results in the incumbent retailer enjoying significant market power over a large portion of its customer base.

In Norway, despite evidence of vigorous competition among the electricity retailers, the prices charged by some suppliers to customers who

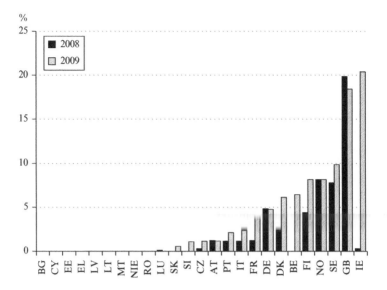

Source: ERGEG 2010 Status Review of the Liberalisation and Implementation of the Energy Regulatory Framework – Ref: C10-URB-34–04.

Figure 6.1   Development of annual switching rates for households (by number of eligible meter points)

have never switched may exceed the best available offers by 10–15 per cent.[12] In the UK, the 2008 Energy Supply Probe[13] found a range of differences in the prices charged to different types of consumers which could not be justified by cost; specifically it found that the incumbent electricity suppliers charged electricity consumers in their former monopoly areas on average over than 10 per cent more than comparable out-of-area customers, and that as many as one-third of switchers may not achieve a price reduction. More recently, in the Retail Market Review,[14] British regulator Ofgem estimated that 40–60 per cent of customers in the energy sector choose not to switch. Furthermore, Ofgem considers that the growing complexity of pricing information discourages switching, as it makes identification of the most favourable offer difficult.

In Italy the regulator recently announced an investigation into the functioning of the retail market, based on preliminary evidence that some consumers supplied by competitive retailers end up paying materially more than the regulated price.[15]

The academic literature echoes the regulators' concerns that electricity retail competition might not deliver the expected benefits to all consumers. Giulietti *et al.*[16] carried out an econometric analysis of the persistence and price dispersion of prices for residential consumers in the UK between 1999 and 2006; they found evidence of limited price convergence as a consequence of firms exploiting significant search and switching costs and creating product differentiation, as well as of significant savings opportunities from switching suppliers that were unexploited by consumers. Wilson and Waddams-Price[17] measured the ability of consumers to choose efficiently between alternative suppliers, based on UK consumer survey data collected in 2000 and 2005. They found that the subsets of consumers who claimed to be switching exclusively for price reasons achieved less than 40 per cent of the maximum gains available through their choice of new supplier, and that more than 25 per cent of consumers actually reduced their surplus as a result of switching. Their analysis also leads to the conclusion that consumers' inefficient choices were not the result of the misleading influence of suppliers' marketing activity. Rather, the results are consistent purely with consumer decision error, possibly related to difficulties comparing complex tariff options.

The concern that retail competition might not deliver the expected benefits, especially to the small consumer, has led most jurisdictions to maintain some forms of regulatory intervention in retail sales. In Europe, the policy has been cast in terms of universal service:

> Member States shall ensure that all household customers, and, where Member States deem it appropriate, small enterprises . . . enjoy universal service, that is the right to be supplied with electricity of a specified quality within their territory at reasonable, easily and clearly comparable, transparent and non-discriminatory prices. To ensure the provision of universal service, Member States may appoint a supplier of last resort[18]

The supplier of last resort is a technical device to ensure that the supplies to passive consumers – those that do not select a competitive retailer – are not interrupted when retail competition is introduced.[19] Passive customers have usually been left with the former monopoly retailer, although nothing prevents them being assigned to competing retailers via auctions for the last resort service.[20] The directive, however, associates the last resort supplier with a broader set of customer protection objectives, which may have provided the legal basis for the decision by several European countries to maintain price controls long after the opening up of the market.[21] In this context, the regulated tariffs[22] charged by the last resort supplier are meant to protect passive consumers until they are in a position to reap the benefits of competition by engaging in the market.

As of 1 January 2010, in 18 European countries end-user regulated prices are enforced on at least one market segment – households, small businesses, medium-sized to large businesses and energy-intensive industry. In most of these countries customers can freely leave the market and switch back to the regulated supplier. Furthermore, despite the fact that some of those countries have adopted a roadmap towards a target situation without price regulation, no commitments as to when this will happen have been taken, in particular for the residential segment.[23]

Even in the UK, where all price controls were lifted in 2002, regulatory constraints to the retailers' pricing strategies have since been introduced, and further measures are currently under discussion. Since 2009 price discrimination between groups of domestic customers without an objective justification is not permitted.[24] In addition, Ofgem is currently considering requiring all retailers to include a standard option in their selection of offers to residential consumers, in order to facilitate price comparisons. The standard option would have a standing charge set by the regulator, while retailers would be free to set the per kWh price component. As a result the standard offers of the various retailers would vary only by the value of the per-unit rate.[25]

Where regulated tariffs are available, only a small share of consumers switch to competitive retailers. Table 6.1 below shows the share of consumers supplied at regulated prices in the countries where price regulation is still implemented, as of 1 January 2010.

In most countries with end-user regulated prices, the share of eligible customers supplied with regulated prices is more than 80 per cent and close to 100 per cent for the residential segment.[26] This evidence hints at a trade-off between the protection of passive consumers against incumbent suppliers and the development of competition in electricity retailing.

No policy approach addressing this trade-off has thus far been widely recognised as superior. A thorough assessment of the measures available to the public authorities in order to minimise the trade-off between the protection of passive consumers and the development of competition is beyond the scope of this book. We merely note that, provided regulated tariffs reflect the cost borne by the supplier of last resort,[27] one may argue that the lack of switching is just a signal that there is no more efficient way to serve consumers. In such a perspective, the removal of price controls before widespread consumer engagement in the market for electricity has been reached appears to reflect a 'protection of the infant industry' argument, that is, the idea that the benefits delivered by retail competition once consumers learn how to take advantage of it outweigh the social cost of attracting competing retailers in the market, due to the exploitation of passive consumers cost in the initial phase of the liberalisation process.

Table 6.1   *Percentage of eligible customers supplied at regulated electricity prices as of 1 January 2010 compared to July 2008*

	Households		Small businesses		Medium to large businesses		Energy intensive industry	
	2008	2010	2008	2010	2008	2010	2008	2010
Bulgaria	n.a.	100.0	n.a.	100.0	n.a.	98.0	n.a.	*
Croatia	n.a.	100.0	n.a.	*	n.a.	*	n.a.	*
Cyprus	n.a.	100.0	n.a.	100.0	n.a.	100.0	n.a.	100.0
Denmark	n.a.	94.0	n.a.	95.0	n.a.	n.a.	n.a.	n.a.
Estonia	**	**	**	**	**	100.0	90.0	100.0
France	99.0	96.0	82.0	83.0	94.0	94.0	82.0	82.0
Greece	100.0	100.0	100.0	100.0	100.0	100.0	100.0	*
Hungary	100.0	100.0	n.a.	n.a.	*	*	32.0	*
Ireland	99.7	79.8	65.0	52.2	45.0	28.0	*	*
Italy	99.7	91.0	79.2	78.2	*	*	*	*
Latvia	100.0	99.0	100.0	99.0	*	*	*	*
Lithuania	100.0	100.0	100.0	n.a.	*	*	*	*
Netherlands	100.0	100.0	100.0	100.0	*	*	*	*
Poland	100.0	100.0	*	*	*	*	*	*
Portugal	97.2	92.0	95.5	88.0	99.5	39.0	100.0	62.0
Romania	100.0	100.0	99.7	n.a.	87.5	n.a.	79.2	*
Slovak Republic	100.0	100.0	*	100.0	*	*	*	*
Spain	92.0	91.0	65.0	*	n.a.	*	*	*

Note: * No end-user price regulation; ** Closed market.

Source: ERGEG Status Review of End-User Price Regulation as of 1 January 2010, Ref: E10-CEM-34–03.

NOTES

1. For the US, see Energy information Administration (EIA), 2010. *Status of Electricity Restructuring by State*, available at: http://www.eia.gov/cneaf/electricity/page/restructur ing/restructure_elect.html. References for Europe are indicated throughout the chapter.
2. In some markets the distribution system operator is responsible for metering, and metering services are supplied to retailers at regulated tariffs; in others the metering activity is liberalised and retailers can procure metering services in the market or self-supply them.
3. However, some fixed-price contracts also insure consumers against changes in the regu-lated transmission and distribution charges for the life of the contract. In this case the risk is typically borne by the retailer.
4. Typically the maximum and minimum hourly volumes that the consumer is allowed to notify are established in the supply contract.

5. In markets designed in such a way that imbalance costs can be reduced by aggregating many consumption and generation nodes in a single balancing account (see Chapter 2, Section 2.3), small retailers may find it profitable to transfer the price imbalance risk of their clients to a party that specialises in aggregating and managing the imbalances.

6. See Joskow, P.L., 2000. *Transaction Cost Economics and Competition Policy*, available at: http://economics.mit.edu/files/1134; and Defeuilley, C., 2009. 'Retail Competition in Electricity Markets. Theoretical Background, Current Situation, Prospects', *Energy Policy*, **37** (2), 377–86.

7. Von der Fehr, N.M. and Hansen, P.V., 2010. 'Electricity Retailing in Norway', *Energy Journal*, **31** (1), 25–46.

8. See, for example, NERA, 2008. *Innovation in Retail Electricity Markets: The Overlooked Benefit*, available at: http://www.competecoalition.com/files/Study_031908.pdf.

9. Joskow, P. and Tirole, J., 2004. *Retail Electricity Competition*, CSEM W.P. 130, available at: http://escholarship.org/uc/item/2rg3z1np.

10. See, for example, Ofgem, 2011. *Retail Market Review: Intervention to Enhance Liquidity in the GB Power Market*, available at: http://www.ofgem.gov.uk/Markets/RetMkts/rmr/Documents1/Liquidity%20Feb%20Condoc.pdf.

11. See Final report prepared for the European Commission by ECME Consortium in 2010, *The Functioning of Retail Electricity Markets for Consumers in the European Union*, p. 77 (Statistics based on consumer survey), available at: http://ec.europa.eu/consumers/consumer_research/market_studies/docs/retail_electricity_full_study_en.pdf.

12. However, no information is available on the share of the customers paying the higher prices. See von der Fehr and Hansen, 2010. (See n. 7, above).

13. Ofgem, 2008. *Energy Supply Probe – Initial Findings Report*, October, available at: http://www.ofgem.gov.uk/Markets/RetMkts/ensuppro/Documents1/Energy%20Supply%20Probe%20-%20Initial%20Findings%20Report.pdf.

14. Ofgem, 2011. *The Retail Market Review – Findings and Initial Proposals*, Reference 34/11, March, available at: http://www.ofgem.gov.uk/Markets/RetMkts/rmr/Documents1/RMR_FINAL.pdf.

15. Deliberazione, 317/2012/E/com 26 July 2012.

16. Giulietti, M., Otero, J. and Waterson, M., 2007. 'Pricing Behaviour under Competition in the UK Electricity Supply Industry', Warwick Economics Research Paper Series (TWERPS) 790, Department of Economics, University of Warwick, available at: http://wrap.warwick.ac.uk/17/; results along the same lines are found in Giulietti, M., Waterson, M. and Wildenbeest, M.R., 2010, 'Estimation of Search Frictions in the British Electricity Market', Warwick Economics Research Paper Series (TWERPS) 790, Department of Economics, University of Warwick, available at: http://wrap.warwick.ac.uk/3513/.

17. Wilson, C. and Waddams Price, C., 2007. 'Do Consumers Switch to the Best Supplier?', Working Papers 07-6, Centre for Competition Policy, University of East Anglia, available at: http://else.econ.ucl.ac.uk/conferences/consumer-behaviour/wilson.pdf; along the same lines, see also Wilson, C. and Waddam-Price, C., 2005. 'Irrationality in Consumers' Switching Decisions: When More Firms May Mean Less Benefit', EconWPA series, available at: http://ideas.repec.org/p/wpa/wuwpio/0509010.html.

18. Article 3 of Directive 2009/72/EC of the European Parliament and of the Council of 13 July 2009 concerning common rules for the internal market in electricity, and repealing Directive 2003/54/EC, Official Journal of the European Union, 14 August 2009, L 211, 55–93.

19. A similar situation occurs for non-passive consumers when a supplier goes bankrupt, until its former clients find a new supplier.

20. This happens for example in Italy. The effects of assigning passive consumers to competing retailers via auctions for the 'default supplier' service are investigated by Bertoletti, P. and Poletti, C., 2012, 'Debiasing through Auction? Inertia in the Liberalisation of Retail Markets', Working Paper no. 47, IEFE Bocconi University

Milan, January, available at: http://economia.unipv.it/pagp/pagine_personali/pberto/papers/deb.pdf.
21. Also along the same lines is the Communication from the Commission to the Council and the European Parliament. Prospects for the internal gas and electricity market, COM 841 final, 2006. The European Commission states: 'Well targeted universal and public obligations, including proportionate price regulation, must remain an integral part of the market opening process . . . Many Member States have retained controls on end-user prices. Although price controls prevent suitable price signals being given to customers about future costs, targeted price regulation may be needed to protect consumers in certain specific circumstances, for instance in the transition period towards effective competition. They must be balanced so as not to prevent market opening, create discrimination among EU energy suppliers, reinforce distortions of competition or restrict resale' (p. 20).
22. Or market-based prices resulting from an auction (see note 21).
23. ERGEG, 2010. *Status Review of End-User Price Regulation as of 1 January 2010*, Ref: E10-CEM-34-03, 8 September, available at: http://www.energy-regulators.eu/portal/page/portal/EER_HOME/EER_PUBLICATIONS/CEER_PAPERS/Customers/Tab1/E10-CEM-34-03_price%20regulation_8-Sept-2010.pdf.
24. This provision came with a sunset clause expiring in July 2012, which Ofgem is now proposing to extend to July 2014 (see Ofgem, 2012. *Consultation on the Undue Discrimination Prohibition Standard License Condition*, 24 February, available at: http://www.ofgem.gov.uk/Markets/RetMkts/rmr/Documents1/Undue_Discrimination_Consultation.pdf.
25. Ofgem, 2012. *The Standardised Element of the Standard Tariff under the Retail Market Review*, available at: http://www.ofgem.gov.uk/Markets/RetMkts/rmr/Documents1/Standardised%20element%20consultation.pdf. For a dissenting view, see Littlechild, S., 2012. *Ofgem's Procrustean Bed: A Response to Ofgem's Consultation on its Retail Market Domestic proposals*, available at: http://www.eprg.group.cam.ac.uk/wp-content/uploads/2012/01/Ofgems-Procrustean-Bed-23-Jan-2012.pdf.
26. Ofgem, 2012 (see n. 25). For a dissenting view, see Littlechild, 2012 (see n. 25).
27. After making up for any cost advantage possibly enjoyed by the last resort supplier over the competitive suppliers because of the former's special legal status.

7. Climate change and the future of the liberalised electricity markets

Guido Cervigni

7.1 INTRODUCTION

As noted in Chapter 1, Section 1.10, the main driver of the expected evolution of electricity systems in industrialised countries is the objective to reduce emissions of the main greenhouse gases responsible for the increase in global mean temperature (global warming).

The Kyoto Protocol, an international agreement linked to the United Nations Framework Convention on Climate Change, sets binding targets for reducing greenhouse gas emissions for 37 industrialised countries and the European Community. The reductions amount to an average of 5 per cent over 1990 levels for the 2008–12 period.

In Europe, the Climate and Energy Package passed at the end of 2008[1] set the 20–20–20 targets to be achieved by 2020: greenhouse gas emissions at least 20 per cent below 1990 levels, 20 per cent of energy consumption provided by renewable resources and a 20 per cent reduction of primary energy use against a baseline scenario.

Given that the deployment of renewables in electricity is more cost efficient than in transport and, to a lesser extent, heating, the burden of the total renewable energy target placed on the electricity sector will be greater. Against a 20 per cent renewable target for total energy, the production of electricity from renewable sources in Europe is expected to rise from 21 per cent in 2010 to 33 per cent in 2020.[2] Greenhouse gas emissions related to the power generation sector are expected to decrease by 3.4 per cent by 2030 compared with 2010.[3]

Figure 7.1 illustrates the expected dynamics of renewable production in the biggest European countries. Wind and solar production, 5 per cent of the total in 2010, will account for about 16 per cent of the electricity generated in Europe in 2020. This will be delivered by a variety of support schemes implemented at national level,[4] including:

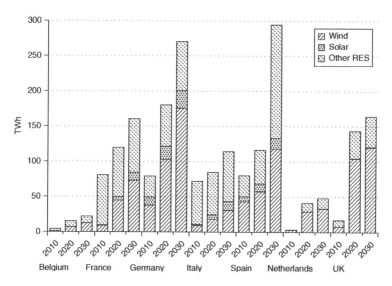

Sources: For year 2010: Eurostat, *Energy Data.* For years 2020 and 2030: European Commission, *EU Energy Trends to 2030.*

Figure 7.1 Expected dynamics of the share of solar and wind generation on total generation from renewable sources

- feed-in tariffs, which pay a regulated price for electricity produced from renewable sources irrespective of the market price for electricity;
- feed-in premiums, uplifts to electricity market prices paid for renewable production; compared with feed-in tariffs these place more risk on the renewable generator, which remains exposed to the volatility of the electricity market price;
- renewable obligations, an obligation to source a certain percentage of electricity supplies from renewable energy sources, typically placed on electricity retailers. The obligation is discharged by surrendering green certificates of renewable production issued by the renewable generators. A market for green certificates is then created, on which renewable generators obtain additional revenues by selling green certificates; and
- tendering systems for the selection of renewable generators admitted to the subsidy scheme – which can take several forms including a feed-in tariff and capacity payments. In this case the size of the subsidy is a by product of the auction process.

A further source of support for renewable electricity generation is the European Union Emissions Trading System (ETS), a cap and trade system on CO_2 emissions introduced in 2005 that covers industrial plants – including power plants over 20 MW, iron and steel plants, along with cement, glass, lime, brick, ceramics, and pulp and paper installations. In addition, aviation is being phased into the ETS, which currently covers almost half the EU's CO_2 emissions. The scheme requires companies to surrender allowances equivalent to their level of CO_2 emissions. A market in carbon allowances is then created, because companies can sell allowances if they cut their own emissions, or buy them if they have insufficient allowances to cover their emissions. The cost of the CO_2 allowances adds to the variable cost of conventional thermal generators, and ultimately to the market price for electricity, which results in additional income for renewable generators that are exposed to the market price for electricity.[5]

Massive transmission investments will be needed in Europe in order to connect the expected 315 GW of renewable capacity expected to be built by 2030. The European Commission has estimated that total investments in European-wide network infrastructures up to 2020 will amount to around €200 billion.[6] The expansion of small-scale generation connected to the distribution grids may lead to power flows from the low-voltage to the medium-voltage grid. Managing bi-directional flows under security conditions will require investments in equipment and possibly significant organisational changes to distribution network operations. The European Commission estimates that by 2020 smart grid investments will reach about €56 billion.[7]

Finally, electricity demand is expected to contribute to the achievement of sustainability objectives. Policy measures to reduce consumption are being implemented in most countries, and further benefits are expected from an increase in the demand response to prices. Enabling small consumers to adjust consumption to changes in spot prices will require massive investment and the development of a large body of new technical, commercial and organisational arrangements.

In Section 7.2 we assess the impact of the growing share of renewable electricity production on the optimal generating capacity mix. We also discuss the new risk allocation between generators and consumers resulting from politicisation of the development of renewable generation capacity. In Section 7.3 we analyse the impact of the growing share of intermittent generation capacity on the markets close to the time of delivery, when the market participants balance their positions and the system operator procures ancillary services. Finally, in Section 7.4 we discuss the possible implications of sustainability objectives on the electricity retail market.

7.2 WHOLESALE ELECTRICITY PRICES AND GENERATION CAPACITY DEVELOPMENT

The recent wave of investment in renewable generation capacity has coincided with a slowdown in the growth of electricity demand brought about by the global economic crisis that began in 2009.[8] The combined effects may have led to excess capacity in some countries such as Italy and Spain in Europe. The end of the nuclear programme in Germany, with the decommissioning of 17 GW of generation capacity by 2020, could mitigate the demand-reducing effect of the economic crisis.

The growing share of renewable capacity will affect the composition of the entire generation fleet in the long run. Figures 7.2 and 7.3 show total load and total load net of wind and solar power in Germany and the UK, respectively, in 2011 for the one thousand hours with the highest load during the year. The figures show that the production profile of renewable

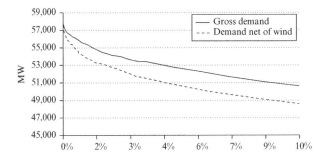

Figure 7.2 *Total load and total load net of wind and solar generation in Germany*

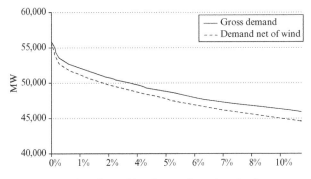

Figure 7.3 *Total load and total load net of wind and solar generation in the UK*

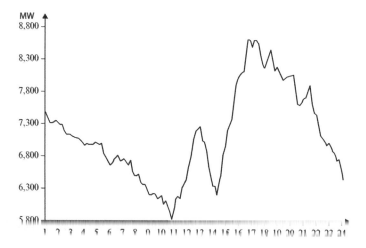

Source: Red Eléctrica de España.

Figure 7.4   Wind production in Spain, 30 March 2008

generators is such that the residual load that needs to be covered by thermal generators is peakier, that is, the proportion of residual load in peak hours is higher than the proportion of total load.

In addition wind production shows great variability. Figure 7.4 reports, for example, wind production in Spain on 30 March 2008. In three hours wind output increased by more than 2,000 MW, about 8 per cent of load.

Wind and solar production is intermittent and poorly predictable far in advance of real time. Figure 7.5 shows the prediction error of wind production in Germany at various times before the time of delivery. The predictions formulated the day before delivery – the traditional timeframe for thermal unit commitment decisions – are highly inaccurate compared with those taken 1–4 hours before real time. Since reliable information on the share of load covered by wind production only becomes available near the time of delivery, an increasing share of non-renewable generation capacity needs to be scheduled, and possibly committed, shortly before real time.

The available evidence suggests that, as the share of renewable generation increases, an efficient generation fleet will include a larger share of units capable of operating efficiently over a wide range of production levels, with fast start-up capabilities and high ramp rates.[9] Other things being equal, average market prices should rise as a result of peaking units being the price makers for a larger number of hours.

Greater renewable production is also expected to result in larger power

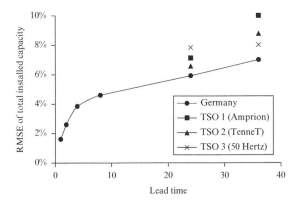

Source: Borggrefe, F. and Neuhoff, K., 2011. 'Balancing and Intraday Market Design: Options for Wind Integration', DIW Discussion Papers 1162, Berlin.

Figure 7.5   *Increasing prediction error of wind generation as a function of the forecast horizon for Germany and three transmission zones*

flows on transmission networks, including across national borders, as renewable primary sources, especially wind, are available only in certain areas. This has recently led to criticism of the congestion management system implemented in Germany and between Germany and Austria, as it results in large-scale unplanned flows through Poland, the Czech Republic, Slovakia and Hungary, which are not the outcome of any market mechanism.[10]

The pursuit of sustainability objectives is leading to a major change in the logic driving investment decisions in generation capacity. The level of installed generation capacity and its composition are ceasing to be the result of decisions taken by market investors, which bear the corresponding risk. Although with different nuances in different countries, the level of renewable capacity is set by political decisions; in some cases, public planning activity also covers the location of new capacity and the financial support schemes are not technology neutral.

Consistently with a planned approach, the bulk of the risk of investment in renewable generation capacity is placed on the final customers. The widely used feed-in tariff schemes make renewable generators' income independent from the market price of electricity.

In this context competition can still play a role, provided that the planning process is supported by and implemented through auctions. This requires the entity with planning responsibility to act on behalf of consumers as the single buyer of renewable production (and/or capacity). Here a tendering system would have a twofold role. The auctions would support planning activity by extracting information on the cost of alternative renewable strategies from the market, and by coordinating transmission and generation capacity expansion. At the same time, auctions would minimise the cost of achieving renewable targets by selecting the lowest-cost providers of renewable generation capacity (and production). Auction-based planning solutions are currently being discussed in some countries,[11] but we are not aware of any fully fledged planning methodology supported by auctions having been implemented yet.

Incidentally, we note that the politicisation of renewable capacity development, and possibly of nuclear production, has an impact on the level and nature of the risk borne by investors in non-renewable generation capacity. Additional regulatory risk is created by the possibility that politically set renewable targets may change, which could have a dramatic impact on the profitability of non-renewable units. This could lead to higher rates of return being necessary to attract investment in generation capacity, and to further pressure on passing on risk to the customers, typically via capacity support schemes such as those discussed in Chapter 3.

7.3   SHORT-TERM AND ANCILLARY SERVICE MARKETS

Renewable energy sources are intermittent and their availability can be accurately predicted only a short time before the time of delivery. As a consequence, the share of total load that needs to be matched by non-renewable generators is only known with certainty near real time. This means that the production programmes of a large share of the generation capacity may have to be modified as the expectations on renewable production are updated.

In traditional hydropower and thermal systems, major changes in the production schedules set the day before delivery have been relatively rare and have been mainly the response to unexpected outages of large units, while demand prediction errors would typically require limited adjustments. However, with a large share of renewable capacity, large swings in production between different units is becoming a normal feature.

Changes in generators' production programmes need to be matched by economic transactions. In the European context these transactions can take place in the intraday or real-time markets. Intraday market transactions take place between market participants. A wind generator that has sold 100 MW in the day-ahead market and realises a few hours before delivery that it will only be able to produce 70 MW, buys the missing 30

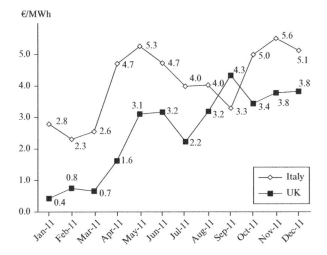

Source: For Italy: Terna. For the UK: National Grid. Monthly Balancing Services Summary.

Figure 7.6   System operators' ancillary services costs in Italy and the UK

MW on the intraday market. Alternatively, the missing 30 MW injections can be supplied by the system operator, which procures it on the balancing market. In this case the wind generator incurs a 30 MW imbalance, which is settled financially with the system operator.

Close to real-time exchanges are increasing in Europe. For example, intraday volumes exchanged in Germany rose from 5.66 TWh in 2009 to 10.3 TWh in 2010, mainly due to the sale of renewable energy.[12] In Spain, intraday volumes doubled between 2008 and 2011, in parallel with the development of a large wind generation capacity.[13] Concerns about increasing system operation costs are also widespread. Figure 7.6 shows the development of system operation costs in Italy and the UK between January 2011 and April 2012.

Moving generation scheduling decisions closer to real time may turn out to be particularly difficult in Europe because of some features of the market design. As we discussed in Chapter 2, in Europe the products traded in day-ahead and intraday timeframes are highly standardised. The standardisation concerns both the location and the time profile of the electricity exchanged. With respect to location, the electricity seller can honour its commitment to deliver by producing at any node of the network within a large control area, which in most cases is the country's borders. The same holds for the buyer. In other words, production and consumption at all nodes of a country's transmission network are regarded as identical products.

With respect to the time profile, the seller of certain volume of electricity for delivery during a certain balancing interval – typically a fixed hour or half-hour – can honour its commitment by producing any quantity at any time during that balancing period, as long as the production over the entire balancing interval adds up to the volume sold. Again, the same holds for the buyer. This means that production and consumption with different time patterns within a balancing interval are regarded as identical products.

As a result, market participants exercising the right to deliver the electricity exchanged at any location and with any time profile may violate system security constraints. These are addressed by the system operator through further transactions on the ancillary service markets, which are run independently from the traded markets. When it comes to relieving congestion, however, production at different locations is not a perfect substitute. In the same way, when it comes to offsetting system imbalances on a second-by-second basis, different time patterns of injections are not equivalent. Consequently transactions in the ancillary service markets take place at different prices depending on the location and time profiles of the electricity bought and sold by the system operator. In other words, little or no product standardisation is implemented on the ancillary service markets.

Figure 7.7 illustrates how these features of the European market design impact on the price formation mechanism. Several products that feature complex substitutability and complementarity are traded close to the time of delivery in separate venues. First, market participants exchange standard

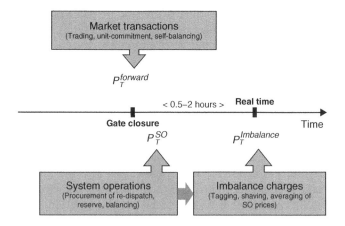

Figure 7.7   Price formation near real time on the European markets

products on the intraday markets. These exchanges create production and consumption commitments, but leave each market participant free to select where within the control area and with which time pattern within the balancing period production and consumption will take place. A by product of product standardisation is that a unique market clearing price applies for deliveries anywhere in the control area during each balancing period.

Second, after market participants' production and consumption programmes have been notified, in some cases while the intraday market is still running, the system operator enters into transactions on the redispatch and real-time markets in order to ensure that all the security constraints are met. These transactions commit generators to increase or reduce production at specific nodes of the transmission network, and with a specific time pattern during the balancing interval. Consequently, different prices apply to electricity delivered at different network locations and with different time patterns.

Finally, the imbalances (deviations between the production and consumption commitments and actual production and consumption) are settled financially between the market participants and system operator, based on metering data. Imbalance prices are related to but are not the same as the prices paid by the system operator on the real-time market. This is because the imbalance prices must be consistent with the commitments made by market participants, as we argued in Chapter 2, Section 2.1. This requires imbalance prices to be location independent and charged on the net imbalance over the entire balancing period.

The co-existence of multiple products with different degrees of standardisation, negotiated independently on the traded and ancillary service markets, may be viable under certain conditions. If the security-constrained generation dispatch does not depart too far and too frequently from the traded market outcome, the prices of standardised products are representative of real demand and supply conditions and end up driving the bulk of production decisions.

In this case, running the intraday and ancillary service markets separately may not bring about major inefficiencies, since significant differences between the clearing prices of those markets only rarely occur. In addition, if demand and supply conditions are relatively stable, and therefore predictable, market participants are able to exploit any arbitrage opportunities between the different markets and products by implementing suitable bidding strategies. Successful arbitrage across the markets results in consistent clearing prices.

Even the administrative constraints implemented in some markets to reduce system operation costs, such as the right for the system operator to disallow intraday market transactions that would create congestion, may cause little harm if they are rarely enforced and if little trading takes place close to real time.

However, the conditions that make segregation of the energy and ancillary service markets viable might not continue to hold as renewable generation production expands. With a large intermittent production capacity, the supply conditions are known only near the time of delivery. More transactions take place close to the real time at prices that may depart significantly from those clearing the day-ahead market.

Furthermore, if the currently observed trends continue in the future, network congestion will become a recurring feature, and the need for fast-response generation capacity will increase. This means that the prices resulting on the re-dispatch and balancing markets – which reflect the real demand and supply conditions – will increasingly depart from the prices of standard products exchanged on the traded markets.

In such a context it may not be possible to achieve consistency between the clearing prices of markets separately run near real time. In addition, the increased need for re-dispatch increases total supply costs, which are ultimately borne by the consumers. This could lead to greater reliance on regulatory and administrative measures aimed at limiting system operation costs, with distortive effects on the prices prevailing in all markets operating close to real time.

The US market design discussed in Section 2.4 can be expected to deal more effectively with the expansion of renewable production. The US standard design is such that the markets for energy and ancillary services are integrated; this ensures the consistency of the prices of all related products. Furthermore, since the same product design is implemented in all timeframes, from the day ahead to real time, the prices that clear the forward and real-time markets differ only to the extent to which the underlying demand and supply conditions have changed. This increases the predictability of the market outcome, and reduces the risk for market participants of missing profitable sales or selling at prices lower than the market-clearing level. Finally, in the US approach, the set of transactions that the market participants can carry out is constrained by the system security constraints. This minimises total supply costs, and makes the introduction of distortive regulatory measures unnecessary.

7.4   THE RETAIL MARKET

The Climate and Energy Package requires primary energy use in Europe to be reduced by 20 per cent compared with a baseline scenario. A broad set of energy conservation policies has been adopted by the European Union over the last two decades, some of which aim to reduce electricity consumption.[14] It is generally thought that greater electricity demand response to prices is also important in order to obtain further benefits. If load reduces when the prices increases, production by peaking generators – which involves higher unit consumption and emissions – falls. An additional benefit of peak reduction, other things being equal, is that less generation and possibly less transmission capacity is needed to meet demand.

Currently a large portion of electricity demand is completely price insensitive in the day-ahead to delivery timeframe. This reflects the technical features of the metering systems currently in place. The meters installed at small consumers' premises traditionally record only total withdrawal since the meter was activated, which allows only calculation of the consumer's withdrawal between two readings. At best, small consumers' meters currently record total consumption over long time periods, such as a month. However, it is impossible to record a consumer's withdrawal per hour. When hourly consumption is not known, retail prices cannot directly reflect wholesale market prices, and therefore cannot convey the economic signals of the value of consumption at each time to consumers.

Business consumers are generally metered hourly and can therefore be charged the hourly market price for electricity. However, supply contracts linked to hourly wholesale prices are not common.

Only some major industrial consumers are currently able to respond to real-time prices; these consumers participate in the spot market and, more importantly, the ancillary service markets, as providers of rapid-response demand reduction capabilities.

Interruptible supply contracts allow the system operator to disconnect consumers when system security is threatened. Typically, interruptible service agreements establish the maximum number of disconnections that the consumer can provide per year, the notice period and the maximum length of each curtailment. Despite providing demand-side flexibility, the load reduction implemented through interruptible contracts is not price driven. Rather, interruptible contracts are exercised by the system operator in emergency situations as a last resort reserve before implementing rolling blackouts. Interruptible contracts are typically offered to industrial and large commercial consumers.

In Europe electricity meters with advanced capabilities, or smart meters will have to be deployed for all consumers by 2020.[15] Recent cost–benefit analyses of the smart meter roll-out to household customers assume that the smart meters will promote a large reduction in electricity consumption, ranging from around 3 per cent (the UK and the Netherlands)[16] to 6 per cent (Ireland) per year.[17] However, these predictions about the reduction of electricity consumption following the introduction of smart meters appear to be somewhat overoptimistic. It is not clear whether such savings will actually be achieved or what contribution will be provided by the smart metering systems. Concerns have been expressed by the British National Audit Office, for example, which considers that the evidence supporting the assessment of electricity savings carried out in the UK is 'inconclusive'.[18]

The development of electric transport could significantly expand the potential for demand-side response, by placing relatively large electricity storage systems on consumer premises. Assuming that the necessary grid investment is carried out, in 2020 electric vehicles are expected to represent a significant share of the vehicle stock in Europe, ranging from 45 to 74 per cent of passenger car and light-duty vehicles.[19]

Many aspects of the technical, commercial and organisational arrangements enabling small consumers to respond to prices are still undetermined. There may be little willingness among small consumers to spend time and resources setting up the systems to monitor and react to spot and real-time electricity prices. In this case retailers need to mediate the increased involvement of the consumers in the market. The scope of retailing activity is therefore likely to widen. In one possible scenario, retailers will operate as dispatchers of their consumers' controllable load and distributed generation capacity. The consumers will sell their suppliers their capability to vary consumption in exchange for lower supply prices, for example by allowing suppliers to remotely activate or deactivate some appliances under predefined conditions.[20] Competition among retailers will determine how the value of the demand response is split between them and their clients.

This sketchy description of a possible scenario, where the price-response potential of small electricity consumers is exploited, suggests that its implementation requires large investments in communication systems, electrical appliances and operational and contractual arrangements. Whether the net benefit of a development in that direction is positive ultimately depends on the actual short-term price elasticity of consumption by small clients, which is as yet largely untested. However, given the complexity and scope of the changes involved, very long development and implementation times are to be expected before the full demand-response potential of small consumers can be realised.

Finally, the value of electricity for the consumers is generally regarded as much greater than the market prices in normal conditions. If this assessment turns out to be correct, the demand response will materialise mainly at times of very high prices. Hence, a necessary condition to obtain benefits from the price elasticity of demand is that no regulatory measures or market design flaws should prevent prices from rising in situations of scarcity.

NOTES

1. The Climate and Energy Package includes several directives and decisions: the Renewable Energy Directive (2009/28/EC), the EU ETS Amending Directive (2009/29/EC), the Fuel Quality Directive (2009/30/EC), the Carbon Capture and Storage Directive (2009/31/EC), and the Effort Sharing Decision (406/2009/EC). For more details, see http://ec.europa.eu/clima/policies/package/index_en.htm.
2. European Commission, 2009. Directorate-General for Energy. *EU Energy Trends to 2030*, available at: http://ec.europa.eu/energy/observatory/trends_2030/doc/trends_to_2030_update_2009.pdf. The reference scenario has been used to collect data on generation. According to the European Commission, the reference scenario assumes that national targets under Directive 2009/28/EC and Decision 2009/406/EC will be achieved in 2020.
3. European Commission, 2009 (see n. 2).
4. For a survey of renewable generation support schemes, see Meyer, N.I, 2003. 'European Schemes for Promoting Renewables in Liberalised Markets', *Energy Policy* **31**(7), 665–76. A description of the schemes implemented in European countries is provided in Canton, J. and Lindén, A.J., 2010. *Support Schemes for Renewable Electricity in the EU*, European Commission, Directorate-General for Economic and Financial Affairs, available at: http://ec.europa.eu/economy_finance/publications/economic_paper/2010/pdf/ecp408_en.pdf.
5. ETS system is currently under review and new arrangements are expected to start in 2013. See http://ec.europa.eu/clima/policies/ets/index_en.htm.
6. European Commission, 2011. *Proposal for a Regulation of the European Parliament and of the Council on Guidelines for Trans-European Energy Infrastructure*, COM/211/658, available at: http://eur-lex.europa.eu/LexUriServ/LexUriServ.do?uri=COM:2011:0658:FIN:EN:PDF.
7. Joint Research Centre of the European Commission, 2011. *Smart Grid Projects in Europe: Lessons Learned and Current Developments*, available at: http://ses.jrc.ec.europa.eu/sites/ses/files/documents/smart_grid_projects_in_europe.pdf.
8. Electricity consumption in EU27 decreased by about 5 per cent between 2007 and 2009. Eurostat, *Energy Database*, available at: http://epp.eurostat.ec.europa.eu/portal/page/portal/energy/data/database.
9. Green, R. and Vasilakos, N., 2009. 'The Long-term Impact of Wind Power on Electricity Prices and Generating Capacity', Department of Economics Discussion Paper no. 11/2009, University of Birmingham, available at: http://www.bhamlive2.bham.ac.uk/Documents/college-social-sciences/business/economics/2010-papers/economics-papers-2011/economics-papers-2011/11-09.pdf.
10. 'Position of ČEPS, MAVIR, PSE Operator and SEPS Regarding the Issue of Bidding Zones Definition', March 2012, available at: http://www.mavir.hu/c/document_library/get_file?uuid=513b0eee-8eb1-405b-85f1-3df85c47237d&groupId=10262.
11. For example, Italy (Legislative Decree no. 28/2011) and the UK (Electricity Market Reform White Paper 2011).
12. Federal Network Agency for Electricity, Gas, Telecommunications, Post and Railways (Bundesnetzagentur). *Monitoring Report 2010*, available at: http://www.bundesnetzagentur.de/SharedDocs/Downloads/EN/BNetzA/PressSection/ReportsPublications/2010/MonitoringReport2010pdf.pdf?__blob=publicationFile.
13. Réd Electrica de España. *Informe del sistema eléctrico 2009, 2010*, available at: http://www.ree.es/sistema_electrico/informeSEE.asp.

14. See 'Energy Savings 2020. How to Triple the Impact of Energy Saving Polices in Europe', available at: http://www.roadmap2050.eu/attachments/files/1EnergySavings2020-FullReport.pdf.
15. Directive 2009/72/EC of the European Parliament and of the Council of 13 July 2009 concerning common rules for the internal market in electricity and repealing Directive 2003/54/EC.
16. DECC & Ofgem, 2011. 'Impact Assessment: Smart Meter Rollout for the Domestic Sector', available at: http://www.decc.gov.uk/assets/decc/11/consultation/smart-metering-imp-prog/4906-smart-meter-rollout-domestic-ia-response.pdf. For the Netherlands: KEMA, 2010. 'Smart Meters in The Netherlands', available at: http://www.rijksoverheid.nl/bestanden/documenten-en-publicaties/rapporten/2010/10/25/smart-meters-in-the-netherlands/10-1193-final-report-smart-metering-ez-draft-v1.pdf.
17. CER, 2011. 'Cost–Benefit Analysis (CBA) for a National Electricity Smart Metering Rollout in Ireland', available at: http://www.cer.ie/GetAttachment.aspx?id=64b7c398-b242-4966-947f-26490b18f117.
18. National Audit Office, 2011. Department of Energy and Climate Change. Preparations for the Roll-out of Smart Meters, available at: http://www.nao.org.uk/publica tions/1012/smart_meterrollout.pdf
19. European Commission, Directorate General for Mobility and Transport, 2011. Study on Clean Transport System, available at: http://ec.europa.eu/transport/urban/studies/doc/2011-11-clean-transport-systems.pdf.
20. For an illustration of a recent demand-side management programme, see, for example, Charles River Associates, 2005. Primer on Demand-Side Management, available at: http://siteresources.worldbank.org/INTENERGY/Resources/PrimeronDemand-SideManagement.pdf.

Index

The economics of electricity markets